A Programmed
Introduction to
Infrared
Spectroscopy

A Programmed Introduction to Infrared Spectroscopy

by B. W. Cook and K. Jones

Imperial Chemical Industries Limited
Petrochemical and Polymer Laboratory

HEYDEN & SON LTD

London · New York · Rheine

Heyden & Son Ltd., Spectrum House, Alderton Crescent, London NW4 3XX.
Heyden & Son Inc., 225 Park Avenue, New York, N.Y. 10017, U.S.A.
Heyden & Son GmbH, Münsterstrasse 22, 4440 Rheine/Westf., Germany.

ISBN 0 85501 032 0 (paperback)
ISBN 0 85501 036 3 (cloth)

Printed in Great Britain by Lowe & Brydone (Printers) Ltd., London

Contents

Introduction

INFRARED SPECTROSCOPY is an analytical technique which finds application in most chemical laboratories. The popularity of this technique is such that even the most junior laboratory assistant is expected to prepare and run samples after receiving the minimum of instruction. Although the instrument manufacturers produce spectrophotometers capable of producing recognisable and presentable spectra with little adjustment to the controls, the chemist is still obliged to prepare the samples. It is this particular aspect which new assistants find the most difficult to master because it is as diverse as the samples they are handling.

The contact between specialist staff in an infrared laboratory and newcomers is often limited. This situation has created the need for a system of basic instruction which enables the relative beginner to study the technique at his own pace. We have produced *A Programmed Introduction to Infrared Spectroscopy* with this purpose in mind. Using the modern programmed teaching method of this book, the reader will be able to assimilate the technique by taking an active part in the learning process, working in his own time at his own speed. Obviously, a really comprehensive knowledge of the subject will only be derived from text books and practical experience.

We gratefully acknowledge the help and encouragement of our colleagues in I.C.I. especially Miss Annette Ramsden and Mr. J. B. Pattison in the compilation of this programme.

How to use this book
You work through the text at your own speed, taking an active part in the learning process. On every page you can prove to yourself that you have learned the material presented; usually a single fact or principle at a time. Having read the material on each page, you will be asked a question and invited to select the correct answer from several choices provided. Each choice is accompanied by a page reference and you turn to the page indicated for the answer you choose. If you choose the correct answer, the new page will confirm it and then go on to present new material. If you choose an incorrect answer, the

new page will correct your mistake and you then go back to the question page to make another choice.

Ideally, you could go through the entire book without choosing a wrong answer, but there is no need to worry if you do make a mistake. Any mistake which you do make will be corrected immediately and you will learn the correct answer before tackling any new material.

The pages are numbered consecutively, but the information is not presented in sequence; for instance you may find the answer page before or after the question page. This is to prevent you from anticipating where the correct answer page may be found. The important thing to remember is to work through the book strictly according to the instructions given on each page. However, if you should wish to take the argument backwards, you will find the page from which you came indicated at the top of each page on the side opposite the page number.

To make the best use of the book, work in periods of about one hour, making notes as you go along. If possible, finish each period at the end of a distinct part.

You will need pencil, notebook and slide rule or tables of logarithms by your side as you tackle this programme, which starts on page 1.

Objectives

THE AIM of this book is to teach the student the principles of infrared spectroscopy, and its practice and use in a laboratory for routine analytical work.

The student will be able:
1. To understand the terms used in infrared spectroscopy.

2. To prepare samples and correctly run spectra from all types of compounds.

3. To identify and remedy both instrumental and sample faults which could lead to poor quality spectra.

4. To understand the principles of operation of the various components of an infrared spectrometer.

5. To understand the reasons for compounds exhibiting infrared activity.

6. To calculate the thickness of cells by the method of interference fringes.

7. To understand the principles of quantitative estimations using Beer's law and the principle of using internal standards.

8. To use correlation tables correctly.

9. To start building a working knowledge of how to interpret infrared spectra.

Validation Report

THE VALIDATION of this programme was based on a Criterion Test (which will be found on pages xiii to xv). The test was used both as a pre-test to eliminate students who had too great a knowledge of the subject and as a post-test to measure the success of the programme.

The participants who evaluated the book ranged from students studying for 'A' levels at Grammar Schools and National Certificates at Technical colleges (both full and part time students) and some graduates. The locations varied through Cheshire, Durham, Lancashire, Yorkshire and Scotland.

All of the participants sat the test as a pre-test. Those scoring more than 20% were eliminated on the grounds that they already had too great a knowledge of the subject. Those scoring 20% or less in the pre-test worked through the programme. On completion each student re-sat the test. There was no time limit for either the tests or the reading of the book.

On the first validation 35% of the participants achieved a score of 75% or higher and parts of the programme were extended or re-written.

The second validation was carried out using a similar group of participants and 62% achieved a score of 75% or higher. Using the formula

$$\frac{\text{amount learned}}{\text{amount that could possible be learned}} = \frac{\text{mean gain between pre- and post-test scores}}{\text{gain between mean pre-test score and full marks}}$$

the efficiency of the book as a teaching programme was calculated to be 67%. The validation exercise was difficult to control and we were very dependent on the goodwill of the participants, and we would like to thank all those who have taken part in the validation of the text and their supervisors.

The Criterion Test

1. What portion of the electromagnetic spectrum is termed the normal infrared region?

2. What are optical filters used for in infrared spectrometers?

3. Name the parts shown in the schematic diagram of an infrared double beam recording spectrometer.

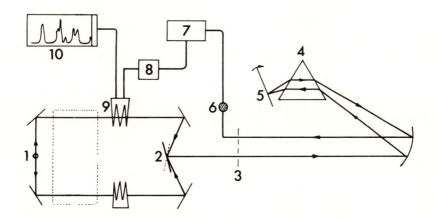

4. Name *three* windows which may be used in the normal 'fingerprint' region of the infrared.

5a. How many modes of vibration has a CO_2 molecule?

5b. How many of these are infrared active?

5c. What property has a vibration to possess if it is to be infrared active?

6a. Name *two* common mulling agents.

6b. Why is it necessary to have *two* different types of mulling agents?

7. Calculate the thickness of the cells which produced the fringe patterns *a* and *b*.

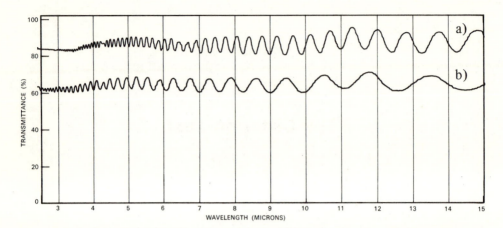

8. What is the formula used to calculate the molar extinction coefficient? Use this to calculate ϵ for the compound $C_6H_{10}O$ given that the cell path length is 47 μm and the concentration is 10% w/v. The spectrum of the absorption band to be used in the calculation is marked x.

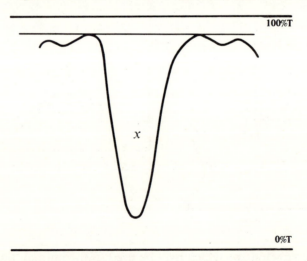

9. Sketch the diagram of the light path in a multi-reflection gas cell, and say why you would use it.

10. What type of table would you use to help you to identify an infrared spectrum of an organic compound?

11. Name *three* different ways of preparing polymer film samples for examination by infrared spectroscopy.

12. What do you understand by (a) fundamental absorption band (b) overtone absorption band (c) combination absorption band?

13a. What is the relationship between wavelength and wavenumber in an infrared spectrum?

13b. What is the wavelength of the infrared radiation equivalent to 1667 cm^{-1}?

14. Between what wavenumbers of the infrared radiation spectrum may you find absorption bands of (a) C=N (b) C=O (c) C−OH?

15a. Name *two* types of detector used in infrared spectrometers.

15b. How does each detector produce its signal?

Check your answers against those given on page 188.

Part One

INTRODUCTION

WHEN YOU sit in front of a fire you feel warm due to the radiant heat travelling out to you from the fire. The radiation which you detect by feeling warm was first examined by Sir William Herschel in 1800. He produced a solar spectrum by placing a glass prism in the path of the sun's rays. He observed the changes which took place, when light of different wavelengths (different colours) fell on the bulb of a sensitive thermometer. He observed that the temperature rose as the thermometer was moved from blue to red but he also found that the thermometer registered even beyond the red end of the visible spectrum. Subsequent experiments showed that this portion beyond the red was composed of a similar type of radiation to visible light, in that it could be reflected, refracted and absorbed by materials which would reflect, refract and absorb visible light.

Question
Why was the radiation from a hot body, which you cannot see, called infrared radiation?

1.   Because infrared is the common name for radiant heat.   → **page 3**

2.   Because if the body is hot enough it appears red and   → **page 5**
      hence the name infrared.

3.   Because this heat radiation occurs at wavelengths
      beyond the red portion of the visible spectrum.   → **page 7**

B

You have chosen the wrong answer. The term infrared means that this part of the spectrum is nearest to the red end of our visible spectrum (the colours we can see with our eyes) but the infrared spectrum is beyond the visible spectrum and cannot be seen with the eye.

In the light of this fact go back to → page 7 and choose a better alternative.

You say correctly that infrared is the common name for radiant heat, but before it became the common name a reason for naming the radiation was required. The question asked why it was so called. You have not answered this question.

Go back to ← page 1 and try again.

You are correct. The nearest visible colour to the infrared in the electro-magnetic spectrum is red.

The wavelength variations in the electromagnetic spectrum are enormous and so it is customary to express the wavelength in different regions of the electromagnetic spectrum in different terms.

The terms used are nanometre (nm), micrometre (μm) and centimetre (cm). It is very useful if you can easily interconvert these units.

$$1 \text{ nm} = 1 \times 10^{-9} \text{ m}$$
$$1 \text{ } \mu\text{m} = 1 \times 10^{-6} \text{ m}$$
$$1 \text{ cm} = 1 \times 10^{-2} \text{ m}$$

e.g. If you wish to express x cm in μm then the calculation you use is as follows:

$$x \text{ cm} = x \times 10^{-2} \text{ m}$$
$$1 \text{ m} = 1 \times 10^{6} \text{ } \mu\text{m}$$
$$x \text{ cm} = x \times 10^{-2} \times 1 \times 10^{6} \text{ } \mu\text{m}$$
$$x \text{ cm} = x \times 10^{4} \text{ } \mu\text{m}$$

If you wish to express y nm in μm you would say:

$$y \text{ nm} = y \times 10^{-9} \text{ m}$$
$$1 \text{ m} = 1 \times 10^{6} \text{ } \mu\text{m}$$
$$y \text{ nm} = y \times 10^{-9} \times 1 \times 10^{6} \text{ } \mu\text{m}$$
$$y \text{ nm} = y \times 10^{-3} \text{ } \mu\text{m}$$

Question
Express 15 μm in nanometres.

1.　　15×10^{-3} nm　　　　　　→ **page 6**

2.　　15×10^{2} nm　　　　　　→ **page 8**

3.　　15×10^{3} nm　　　　　　→ **page 16**

The statement that hot bodies appear red is partially correct. However, the name given to the radiant heat emitted by such a body did not arise because of this phenomenon. You have not thought carefully enough about the question.

Go back to ← page 1 and choose another alternative.

You are wrong. Let us go through the calculation again. If you treat all of the units as portions of a metre and reduce to this state the results you get are as follows:

$$1 \, \mu m = 1 \times 10^{-6} \, m$$
$$= 1/10^6 \, m$$
$$1 \, m = 1 \times 10^9 \, nm$$

If you express $1/10^6$ m as nanometres the equation you get is:

$$\frac{1}{10^6} \, m = \frac{1}{10^6} \times 10^9 \, nm$$

Now work out what 15 μm are in nanometres.

In the light of your answer, go back to ← page 4 and choose the correct alternative.

You are correct. Infrared means beyond the red, so this is the portion of a spectrum where this radiation is found.

The existence of many kinds of radiation have now been discovered. The radiations all travel at the same speed and differ only by their wavelengths. The visible and infrared regions form only a small portion of this system which is known as the electromagnetic spectrum (Fig. 1.1).

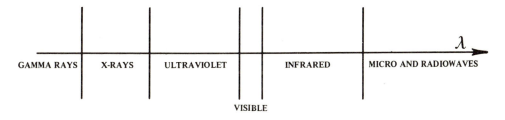

Fig. 1.1

Question
What visible colour is nearest to the infrared region in the electromagnetic spectrum?

1. Blue   ← page 2

2. Red   ← page 4

3. Green   → page 10

You are wrong. Let us go through the calculation again. If you treat all of the units as portions of a metre and reduce to this state the results you get are as follows:

$$1 \, \mu m = 1 \times 10^{-6} \, m$$
$$= 1/10^6 \, m$$
$$1 \, m = 1 \times 10^9 \, nm$$

If you express $1/10^6$ m as nanometres the equation you get is:

$$\frac{1}{10^6} m = \frac{1}{10^6} \times 10^9 \, nm$$

Now work out what 15 μm are in nanometres.

In the light of your answer, go back to ← page 4 and choose the correct alternative.

Your answer is wrong. The wavelength represented by 10^{-3} cm is calculated as follows:

$$10^{-3} \text{ cm} = 1/1000 \text{ cm}$$

What fraction of a metre is 1/1000 cm?

In the light of your answer, go back to → page 16 and choose another alternative.

You have chosen the wrong answer. The term infrared means that this part of the spectrum is nearest to the red end of our visible spectrum (the colours we can see with our eyes) but the infrared spectrum is beyond the visible spectrum and cannot be seen with the eye.

In the light of this fact go back to ← page 7 and choose a better alternative.

Correct. 10^{-3} cm is the same as $10\,\mu m$.

There is one other term used by spectroscopists when describing positions of absorption bands in the infrared and it is the wavenumber of the absorption. The wavenumber is the number of electromagnetic waves of wavelength λ which will occur in 1 cm.

Do you remember that on page 4 we found that 1 cm was equivalent to $1 \times 10^4\,\mu m$? The number of waves in 1 cm can be found by dividing the length (1 cm) by the wavelength of the radiation
e.g.

$$\frac{1\ cm}{\text{wavelength } (\lambda)} = \frac{1 \times 10^4}{\lambda}$$

We have to write $1 \times 10^4\,\mu m = 1$ cm because the wavelength λ is expressed in micrometres. The wavenumber is written as cm^{-1} because it is the number of waves per centimetre.
e.g.

$$10\,\mu m \equiv 1/10 \times 10^4\ cm^{-1}$$
$$\equiv 1000\ cm^{-1}$$

From the above expressions it is shown that the wavenumber is equal to the reciprocal of the wavelength multiplied by 1 cm expressed in micrometres
i.e.

$$cm^{-1} = 1/\lambda \times 10^4$$

Question
How would you express $5\,\mu m$ in wavenumbers?

1. $1000\ cm^{-1}$ → page 15

2. $2000\ cm^{-1}$ → page 17

3. $500\ cm^{-1}$ → page 13

1. The colour observed for a compound with maximum absorption at 790 nm would be **green**.

2. The colour observed for a compound with maximum absorption at 20000 cm^{-1} would be **red**.

Were you correct and do you know how these answers were derived? If you were wrong and wish to see how they were derived, turn to **page 14**.

If you wish to continue with the programme Part 2 starts on → page 19.

I am sorry you are wrong. The wavenumber scale is the reciprocal of the wavelength expressed in centimetres.

Express 5 μm in centimetres (remember 1 μm $\equiv$ 10^{-6} m) and take the reciprocal of this number.

In the light of your answer and the statement above, go back to ← **page 11 and choose the correct alternative.**

1. The question asked which colour which would be observed for a compound with an absorption maximum at 790 nm.

790 nm is a red colour, and because it is absorbing at this wavelength the colour transmitted will be the complement of this, which is green. The colour wheel is shown below (Fig. 1.4).

2. Do you remember how to convert wavenumbers into wavelengths? If you do not please turn to ← page 11 and re-read the section.

$$20000 \text{ cm}^{-1} \equiv 500 \text{ nm}$$

500 nm is green and the complement to this is red, so the colour you would observe is red, as defined by the colour wheel shown below (Fig. 1.4).

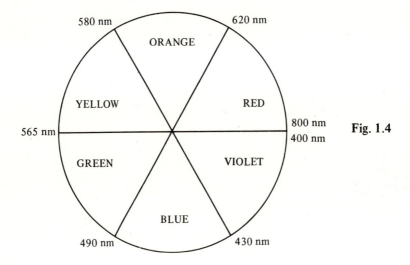

Fig. 1.4

Part 2 starts on → page 19.

I am sorry you are wrong. The wavenumber scale is the reciprocal of the wavelength expressed in centimetres.

Express 5 μm in centimetres (remember 1 μm $\equiv$ 10^{-6} m) and take the reciprocal of this number.

In the light of your answer and the statement above, go back to $\leftarrow$ page 11 and choose the correct alternative.

You are correct. The answer is 15×10^3 nm.

Different terms are used in each region so that the figures used to express the wavelength are of reasonable size and do not have excessive powers of 10 in their constitution.

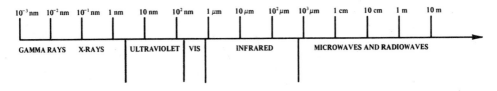

Fig. 1.2

The infrared region of the electromagnetic spectrum (Fig. 1.2) is the region which includes the wavelengths between 0.7 μm and 100 μm. The portion of this region which is the most useful is called the fingerprint region and is bounded by the wavelengths 2.5 μm and 15 μm. Wavelengths in the infrared are expressed always in micrometres (μm).

Question
How would you write the infrared wavelength equivalent of 10^{-3} cm?

1. 100 nm   ← page 9

2. 1 μm   → page 18

3. 10 μm   ← page 11

You are correct. 5 μm is equivalent to 2000 cm^{-1}.

When a source of white light passes through a clear glass plate no change of colour can be seen and we can say that the clear glass plate did not **absorb** any of the light.

If, however, we replace the clear glass plate with a piece of red glass the transmitted light seen is red. Thus the red plate absorbs some of the white light and only red light is transmitted. The colour exhibited by most absorbing materials either by reflection or by transmission will be the complement of the colour absorbed. Qualitatively the colour may be predicted by means of the colour wheel (Fig. 1.3).

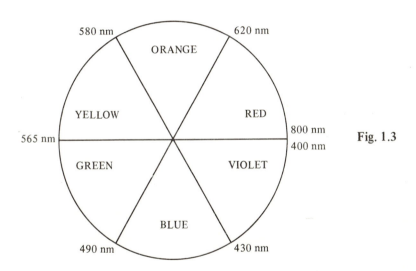

Fig. 1.3

Question
What colour would be observed for:

1. a compound with maximum absorption at 790 nm?

2. a compound with maximum absorption at 20000 cm^{-1}?

Now turn to ← page 12 and check your answers.

Your answer is wrong. The wavelength represented by 10^{-3} cm is calculated as follows:

$$10^{-3} \text{ cm } = \ 1/1000 \text{ cm}$$

What fraction of a metre is 1/1000 cm?

In the light of your answer, go back to ← page 16 and choose another alternative.

Part Two

SPECTROMETER COMPONENTS

WE KNOW from looking at rainbows that what appears to our eyes as white light is, in fact, a mixture of colours, all of which blend together to form white light.

How can we produce this splitting up of light into its **visible spectrum** in the laboratory?

If white light falls onto a glass prism as shown in Fig. 2.1 the prism will refract the white light, and the dispersed radiations will form the visible spectrum, consisting of the colours of the rainbow: red, orange, yellow, green, blue, indigo and violet.

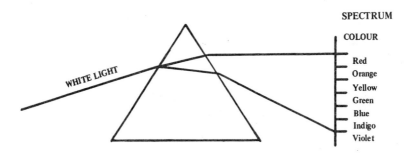

Fig. 2.1

Question
Which of the following wavelengths of white light is refracted the most when passed through a prism?

1. 700 nm   → **page 25**

2. 600 nm   → **page 30**

3. 400 nm   → **page 28**

The first three orders which will be returned along the path are:

1st	20 μm	500 cm^{-1}
2nd	10 μm	1000 cm^{-1}
3rd	6.667 μm	1500 cm^{-1}

Did you arrive at these answers? If so, continue by turning to → page 24. If you made any mistakes and would like the answers explained, turn to → page 22.

Correct. A light bulb has a glass surround to the filament and this absorbs a major amount of the energy given out by the source; therefore the source used is the Nernst filament.

Now we have a source of infrared radiation, the next requirement is a detector. Because infrared radiations are essentially radiant heat, thermal detectors are used to detect changes in the radiations.

Thermal detectors are made as small as possible to reduce their heat capacity so that for a given amount of energy there will be a large temperature rise. This causes a change in the electrical response of the detector which may be amplified and then the signal recorded. In order to make the detector rapid in response it must be able to dissipate the heat very rapidly, that is, it must have a large area to weight ratio. There are three main types of detector.

The Thermocouple. This uses the principle that the change in temperature of a junction of two dissimilar metals or semi-conductors creates an electro-motive force (e.m.f.) which may be measured. The thermocouples are sealed into permanently evacuated envelopes. These envelopes have windows that allow the radiations to enter. This reduces the conduction loss and so increases the sensitivity.

The Bolometer. This uses the principle that the electrical resistance of a pure metal or semi-conductor is temperature-sensitive. If a constant potential is applied to such a detector the variation of the resistance with temperature may be measured by the variations in the current flowing in the circuit.

The Golay cell. In this cell the detector is a small metal cylinder enclosed by a blackened metal plate at one end and by a flexible metallised diaphragm at the other. The cylinder is filled with a gas and sealed. The infrared radiations fall onto the blackened plate and the heat is conducted through to the gas causing it to expand and deform the metallised diaphragm. Light from a lamp inside the detector is focused onto the diaphragm. The light is reflected from the diaphragm and falls onto a photo-electric cell. Movement of the diaphragm moves the light beam across the photocell. The photocell output is directly proportional to the expansion of the gas caused by the infrared radiation falling on the blackened plate.

Question
On what principle does the thermocouple detector operate?

1.   Change in gas expansion   → **page 23**

2.   Change in e.m.f.   → **page 26**

3.   Change in resistance   → **page 32**

When the blaze angle = 30° and the ray is returned along its path, the angle of incidence = the angle of diffraction.

The formula we must use is: $n\lambda = d(\sin\theta + \sin i)$ and, as we have said above, the angles θ, i and β are the same and equal to 30°.

Hence, $n\lambda = 2d \sin\beta$.

You were given that $\beta = 30°$ ($\sin\beta = \frac{1}{2}$) and $d = 500$ facets/cm.

Thus when $n = 1$
$$\lambda = 2 \times 1/500 \times \frac{1}{2}$$
$$= 2 \times 10^{-3} \text{ cm}$$
$$= 20\,\mu\text{m}$$

Now to calculate the wavenumber equivalent to $20\,\mu$m you work out the number of waves per centimetre

$$\text{wavenumber} = 1/20 \times 10^4 \text{ cm}^{-1}$$
$$= 500 \text{ cm}^{-1}$$

Continue by substituting 2 and 3 for n in the formula

$$n\lambda = 2d \sin\beta$$

and you will obtain the following answers:

when $n = 2$ $\lambda = 10\,\mu$m; 1000 cm^{-1}
when $n = 3$ $\lambda = 6.67\,\mu$m; 1500 cm^{-1}

Continue by turning to → page 24.

I'm sorry but you are wrong. A thermocouple does not have a gas reservoir. It is the Golay cell which uses the change in gas expansion to measure changes in infrared radiation.

In the light of this fact go back to ← page 21 and choose a better alternative.

The source of a spectrometer which has to be used in the infrared may be any convenient heat source, that is, any source which has a temperature greater than 1200°C. An electric light bulb may be used in the region 0.5 μm to 4 μm. It is limited to this region because above 4 μm the surrounding glass absorbs the radiation. For the range 0.7 μm to 50 μm the source most commonly used is the Nernst filament.

The Nernst filament is a bar or hollow tube about 5 cm in length and about 2 mm in diameter. It is composed of a mixture of rare earth oxides, similar in composition to the white radiants of a gas fire. The filament operates best from the mains supply of 250 V and between 0.5 and 1 A, depending on the diameter of the filament, and has a surface temperature of between 1500°C and 1800°C.

Question

An electric light bulb is a very stable and rich source of radiation and is used in visible/ultraviolet spectrometers; why is it not used in infrared spectrometers?

1. Because it only gives out visible/ultraviolet   → **page 31**
    radiation.

2. Because the source of the radiation is surrounded   ← **page 21**
    by a material which absorbs a large amount of the
    infrared radiation.

3. Because it does not emit any radiation beyond 4 μm.   → **page 27**

You are wrong. Do you remember that in Fig. 2.1 we showed that blue light was refracted more by a prism that red light? On page 17 we described the colours and their wavelengths. The question you must now answer is which of these wavelengths is nearest to blue?

In the light of your answer and the statements above, go back to ← page 19 and choose the correct alternative.

Correct. The thermocouple uses the principle that the change in e.m.f. at a junction of two different metals may be measured.

Now that we have the basic parts of the spectrometer let us consider the scanning spectrometer.

The part of an infrared spectrometer containing the grating or prism is termed a **Monochromator** and produces what is called **Monochromatic** light. Monochromatic means one (mono) colour (chrome). A monochromator con- sists of an entrance slit S_1, a concave mirror M_1 which reflects the light as a parallel beam onto the prism or grating and a second mirror M_2 which re- focuses the light onto the exit slit S_2 (Fig. 2.5).

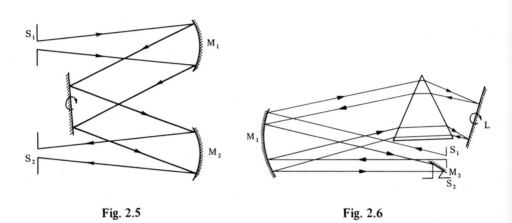

Fig. 2.5   Fig. 2.6

The methods used for scanning the spectrum across the exit slit are different in prism and grating monochromators.

In a prism monochromator it is usual to keep the prism in a fixed position and to turn the Littrow mirror (L) so as to scan the spectrum across the exit slits. As you can see from Fig. 2.6 this causes the beam of radiation to pass through the prism twice. As the beam is dispersed at each passage the total dispersion is double that of a single pass.

In a grating monochromator it is usual to simply rotate the grating and so scan the reflected spectrum across the exit slit (Fig. 2.5).

Question
In which type of monochromator is the scanning produced by turning a mirror?

1.   A grating monochromator   → page 34

2.   A prism monochromator   → page 37

The components of a light bulb are a filament, surrounded by a glass bulb, and it is because of the surrounding glass that it is not used in infrared spectroscopy. Can you say why the glass would interfere with the radiation from the source?

When you have answered this question correctly you will be able to choose the correct alternative on ← page 24.

You are correct. The shortest wavelength of light is refracted the most when it passes through a prism.

Another common method of dispersing light is to use a grating. The type of grating generally used in infrared spectrometers is a reflection grating and consists of a number of adjacent polished facets. When monochromatic light is incident on a grating, beams of light are reflected at a number of angles which correspond to a path difference of $n\lambda$ between rays striking adjacent facets as shown in Fig. 2.2.

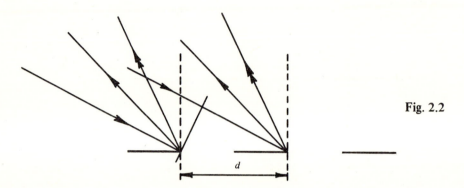

Fig. 2.2

When white light falls on the grating, dispersion of the radiation occurs and several spectra are produced. Each one, however, is not a pure spectrum in that the spectra overlap one another. These are the 1st, 2nd and 3rd 'order' spectra and are brought about by substituting 1,2,3 etc. for n in the formula:

$$n\lambda = d\,(\sin\theta + \sin i)$$

where θ = angle of diffraction and i = angle of incidence, when both are measured in the same direction from the perpendicular and d = distance between facets (Fig. 2.3).

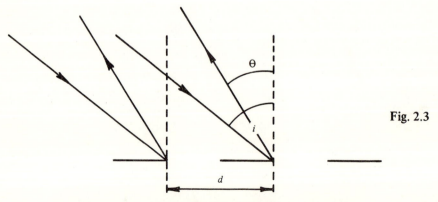

Fig. 2.3

Continue the programme by turning to → page 29.

This means that to obtain a single colour at a particular angle it is necessary to filter out the unwanted colours and so optical filters or a fore prism are used in conjunction with a grating in a spectrometer.

The intensities of the various orders are influenced by the geometry of the facets. One valuable type, and the one used in infrared spectroscopy, is the echelette reflection grating which is ruled as in Fig. 2.4.

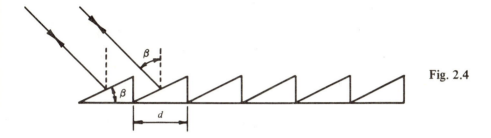

Fig. 2.4

This concentrates nearly all the light incident at the blaze angle β into the 1st order and reflects it back along its path of incidence. The other orders are present but are of a much reduced intensity.

Question
A certain echelette reflection grating has 500 facets/cm and a blaze angle of 30°. Calculate

1. the wavelengths, and

2. the wavenumbers of the light that will be observed at the blaze angle for the first three 'orders'.
   (Assume that the angle of incidence = angle of diffraction).

Now turn to ← **page 20 and check your answers.**

You are wrong. Do you remember that in Fig. 2.1 we showed that blue light was refracted more by a prism than red light? On page 17 we described the colours and their wavelengths. The question you must now answer is which of these wavelengths is nearest to blue?

In the light of your answer and the statements above, go back to ← page 19 and choose the correct alternative.

A light bulb is a source of radiation which is safe for common usage. Because it is a very rich source of these radiations, it is primarily used for visible and ultraviolet spectroscopy. It is also a very hot source. One must therefore conclude that there is another reason why it is not used in infrared spectroscopy.

In the light of these facts go back to ← page 24 and choose a better alternative.

I'm sorry but you are wrong. The change in resistance of a detector is usually used in the bolometer and not the thermocouple.

Re-read ← page 21 and then choose the correct alternative.

You are correct. It is only in a grating monochromator that filters are used.

We have found that infrared light is detected by its heating effect, causing changes in the electrical properties of a detector.

If we just had a simple spectrometer consisting of a Nernst filament, a prism (or grating), focusing mirrors (or lenses) and a thermal detector we would find that every time the sun came out or a bunsen burner was lit the detector would 'see' the increased background signal and its output would change and confuse any readings being taken.

This difficulty can be overcome by placing what we call a 'chopper' close to the source.

A chopper may be a disc of metal with segments removed. This disc or chopper is rotated by an electric motor.

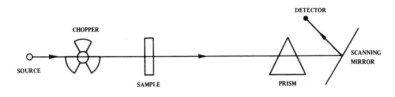

Fig. 2.7

The light from the source alternately passes through the holes in the chopper or is stopped by the solid metal. The detector therefore 'sees' a flashing light and its signal varies from zero to a maximum value (Fig. 2.7).

The amplifier is set to amplify only signals at the frequency of the chopper. Thus any changes of background radiation level will still affect the detector, but since this radiation will not have passed through the chopper it will not be amplified and hence will not affect the readings being taken.

Question
Is a chopper placed close to the source because:

1. it amplifies the signal?   → page 38

2. it increases the signal to a maximum value?   → page 43

3. it eliminates the effects of changes in background   → page 40
    radiation?

D

In a grating monochromator the grating itself is turned to scan the spectrum across the exit slit. Although the grating is a reflection grating it is not used as a mirror in the accepted sense as this would be the zero order spectrum and there would be no dispersion of the light incident on the grating.

Return to ← page 26 and read the text more carefully.

You are wrong. The definition of resolution tells of the ability of a spectrometer to separate adjacent bands. What type of slits increases the resolution of the spectrometer? Are more bands separated in spectrum A than B?

When you have answered these questions correctly you will be able to choose the correct alternative on → page 44.

You are wrong. Their spectra are very different.

The double beam spectrometer is the common infrared spectrometer manufactured today and this is a much more complicated instrument than a single beam spectrometer and therefore it must have advantages.

Do you remember that in the text we said that the spectrum produced by a single beam spectrometer was superimposed onto the emission curve produced by the Nernst? A double beam spectrometer, however, has two beams which are compared. Would you expect this to make a difference?

Re-read the section on single beam and double beam spectrometers on → page 41 and select another alternative.

You are correct. In a prism monochromator the spectrum is scanned across the exit slit by turning the Littrow mirror.

If we use a grating as the monochromator element it is necessary to remove the overlapping orders. This can be done in two ways:

(a) Using filters. These are generally thin discs of a material which only transmit over a short wavelength range, thus only the order corresponding to this wavelength range will be transmitted. It usually requires several filters to cover the full range of the fingerprint region.

(b) Using a prism. The prism may be placed either before or after the grating. This refracts the light so that only the order required passes across the exit slit.

Question
Where would you expect to find optical filters in an infrared spectrometer?

1.   In a grating monochromator.   ← page 33

2.   In a prism monochromator.   → page 45

3.   In a prism/grating monochromator.   → page 42

A chopper is a mechanical device used to eliminate changes in background radiation. It has no amplifying effect at all, hence you have chosen wrongly.

In the light of this information go back to ← page 33 and choose the correct alternative.

A labelled schematic diagram of an infrared spectrometer is shown in Fig. 2.13.

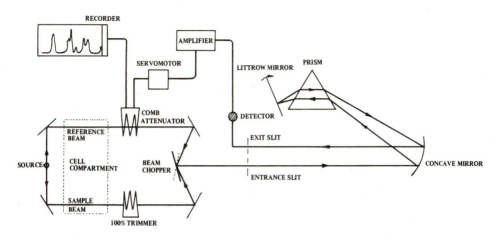

Fig. 2.13

Continue the programme by turning to → page 48.

You are correct. A chopper is used in a single beam spectrometer to eliminate changes caused by stray light to the background level of radiation.

Let us now continue by combining all of the parts together.

Simple bench spectrometers use lenses to focus the light beam. This is not practical in an infrared spectrometer as most suitable lens materials (NaCl and KBr) are water soluble and easily scratched.

There are two distinct types of infrared spectrometer:

Single beam. This is where the light from the source is chopped, focused onto the sample, dispersed and detected. The light detected is deficient in the wavelengths absorbed by the sample. Because the source does not emit evenly at all wavelengths (see Fig. 2.8) the sample spectrum is superimposed onto a changing background.

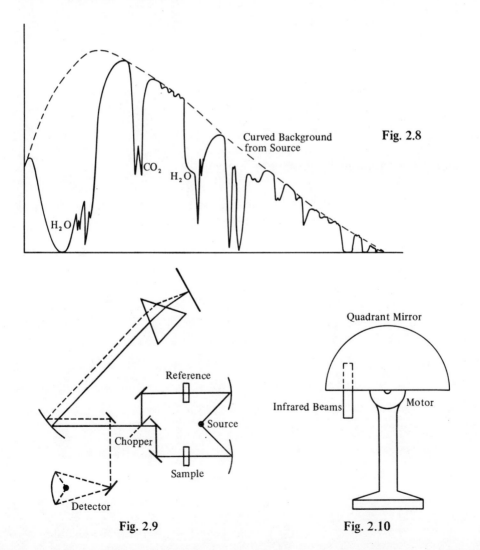

Curved Background from Source

CO_2

H_2O

H_2O

Fig. 2.8

Quadrant Mirror

Reference

Source

Infrared Beams

Motor

Chopper

Sample

Detector

Fig. 2.9

Fig. 2.10

Single beam spectrometers must be kept dry and free from carbon dioxide as water vapour and carbon dioxide absorb infrared radiation. This sort of spectrometer is seldom used nowadays.

Double beam. A double beam spectrometer has two identical light beams taken from the same source (Fig. 2.9). One beam (sample beam) passes through the sample, the other (reference beam) sees only the atmosphere. The chopper is placed after the sample and serves to bring the two beams along the same path through the monochromator and onto the detector. (It also serves the same function as in the single beam spectrometer i.e. removal of stray radiation).

The chopper is composed of rotating or reciprocating mirrors. If the mirror is as shown (Fig. 2.10) and it is set in the spectrometer as above, then the radiation from the reference beam and the sample beam are at the same height, that is, just above the pivot of the chopper. The radiation from the sample beam is reflected into the monochromator when the chopper mirror is on the right side and the reference beam is blocked by the mirror. The rotating mirror means that the detector sees the beams alternately. When there is nothing in the beams the detector cannot measure any difference, because it is seeing a continuum.

As the sample begins to absorb light, the electrical signals at the detector become unequal and an out-of-balance signal arises. This signal is fed to a small servo motor which drives a shutter into the reference beam until there is no out-of-balance signal at the detector.

As the sample absorbs more light the shutter moves further across the reference beam and maintains both sample and reference beams at balance. As the sample absorbs less light the shutter moves out of the reference beam and hence again maintains the beam at balance. The shutter movement is fed either electrically or mechanically to a recorder pen. This pen traces out a graph of the absorption of the sample versus wavelength.

Double beam spectrometers need not be dried because as both beams travel the same distance and are compared at the detector, equal amounts of light are absorbed by the atmosphere.

Question
Is there a difference in the appearance of the spectra from a single beam spectrometer and a double beam spectrometer?

1. Yes, the spectrum from a double beam instrument is twice the intensity of that from a single beam instrument. → page 46

2. Yes, the background from a double beam instrument is flat whilst that from a single beam instrument follows the Nernst emission curve. → page 44

3. No, they are identical. ← page 36

You are wrong. A grating/prism combination is widely used in infrared spectrometers because the prism removes unwanted grating orders and hence filters would serve no useful purpose.

In the light of this information go back to ← page 37 and choose the correct alternative.

I am sorry, you are wrong. The chopper is used to eliminate changes in background radiation. It has no effect on the signal strength.

In the light of this information go back to ← page 33 and choose the correct alternative.

44

You are correct. There is a difference in the appearance of the spectra produced by single and double beam spectrometers. The single beam has a curved background whilst the double beam has a flat background.

In the frame about monochromators (page 26) we saw that they had two slits: the entrance slit and the exit slit. Now let us consider the exit slit.

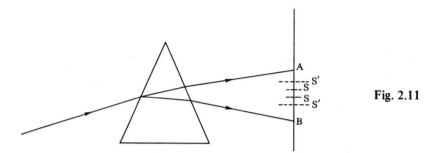

Fig. 2.11

Look at Fig. 2.11. The prism disperses the incident radiation over the area bounded by A and B. The radiation passing through the narrow slits S—S contains a smaller frequency spread than the radiation passing through the wide slits S'—S'. The radiation which passes through the narrow slits is nearer to monochromatic radiation than that passing through the wide slits. Assuming that the energy of the dispersed beam is uniform over its area, it is obvious that more energy will pass through S'—S' than through S—S. Summarising, wide slits give a broad beam of high energy but low resolution. Narrow slits give low spectral energy but high resolution.

We define the **resolution** of a spectrometer by the ability to separate two adjacent sharp bands. The closer the bands that are separated, the better is the resolution of the spectrometer.

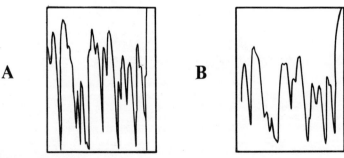

A   **B**

Question
The two spectra A and B are of the same sample but have been run using different instrumental conditions. What was the instrumental condition used for running spectrum A?

1. Wide slits. ← **page 35**

2. Narrow slits. → **page 47**

You are wrong. Filters are used to remove unwanted spectral orders and since a prism gives only one spectral order no filters are required.

In the light of this information go back to ← page 37 and choose the correct alternative.

You are wrong. Only the absorption from the sample can affect the intensity of the spectrum. Thus, changing instruments has no effect on the intensity of the sample's absorption pattern.

Re-read ← page 41 and select the correct alternative.

You are correct, narrow slits were used. This is shown by the fact that the resolution of the bands in spectrum A is better than that in spectrum B.

Figure 2.12 is a schematic diagram of an infrared double beam recording spectrometer. List the names of as many parts as you can.

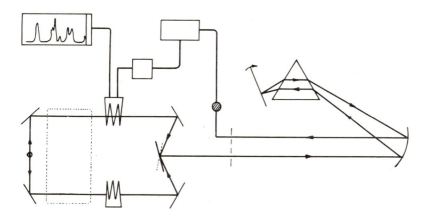

Fig. 2.12

On completion turn to ← page 39 and check your answers.

Now let us look at the parts of an infrared spectrometer which we have learnt about.

A **Nernst filament** is a source of infrared radiation which is focused onto the entrance slit of:

The **monochromator** whose prism's Littrow mirror or grating is turned to produce a beam of monochromatic light passing across the exit slit. This emergent beam is focused onto:

The **detector** whose out-of-balance signal is fed to:

An **amplifier** which electronically produces a larger signal. This is fed to:

A **servo motor** which drives a compensating comb into or out of the reference beam. A signal from this servo motor is also fed to:

A **recorder** whose paper moves in step with the prism or grating movement. Hence the servo motor movement required to maintain the two beams at balance is drawn out on the recorder chart as an infrared absorption spectrum.

The type of result you should obtain from an infrared recording spectrometer is as shown in Fig. 2.14.

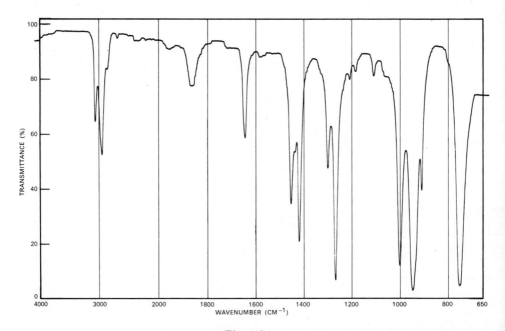

Fig. 2.14

Part 3 starts on → page 49.

Part Three

SETTING UP A SPECTROMETER

In this part the pages are arranged in numerical order, and can be read in the same way as a normal text book.

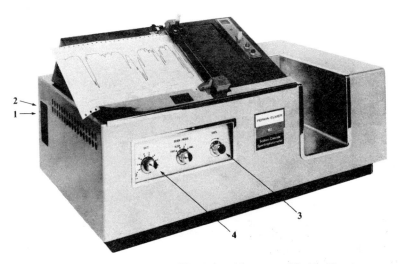

Fig. 3.1 *(Courtesy of Perkin Elmer)*

FIGURE 3.1 shows the layout of a commercial spectrometer, the Perkin Elmer 157. We will go through the controls and describe their functions in the following order:

1. Balance control

2. Gain control

3. 100 per cent control

4. Slit control

50

Balance control

A control of a spectrometer with which we must be conversant, but which needs only infrequent adjustment, is the electrical balance potentiometer.

Let us first look at what would happen if we were to simultaneously blank off both beams of the spectrometer with a thick card. The card does not allow radiation from the source to enter the monochromator. This means that the detector is not giving any signal to be amplified and recorded. This will cause the recorder pen to draw a straight line parallel to the 100 per cent transmission line (Fig. 3.2).

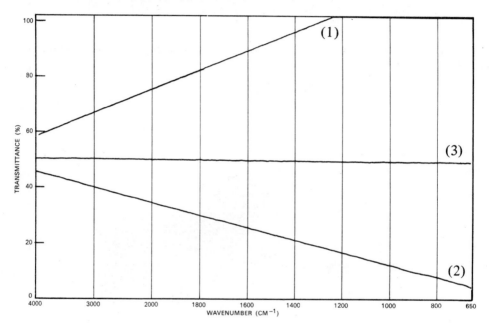

Fig. 3.2   (1) and (2) are incorrect balance lines, (3) is the correct balance line.

If the amplifier is incorrectly set the recorder pen may drift up or down because of the steady out-of-balance signal in the amplifier. If this state exists the electrical balance potentiometer is turned until the drift has been cancelled. This adjustment is usually carried out at intervals of one month.

Gain control

This is an electronic control which adjusts the amount by which the signal fed to the servo motor is amplified. When the gain is high the recorder pen will show excessive 'noise' (unsteadiness) and in extreme cases will oscillate vigorously (Fig. 3.3). When too little gain is being used the pen will respond sluggishly (Fig. 3.4).

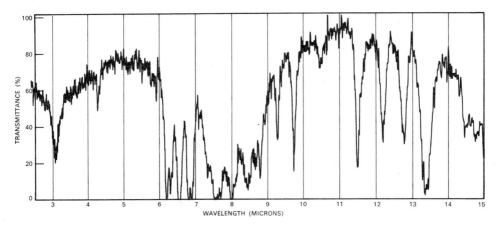

Fig. 3.3

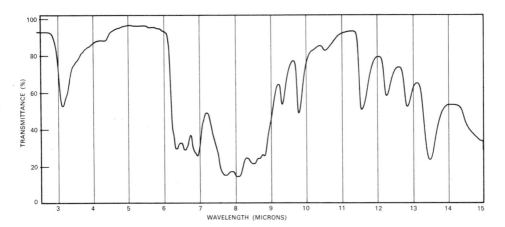

Fig. 3.4

Adjustment of gain. With the sample in place, rapidly scan through until you come to a position in the spectrum where the pen is near the top (100%) of the chart. In the case of a grating instrument this will be approximately half way through the range being used. If it is a prism instrument, find a high point as near as possible to, but on the high frequency side of, 1100 cm^{-1}.

Remove the pen from the paper. Place your hand in the sample beam and then rapidly remove it so that the pen returns to its position of rest. Adjust the gain control so that as the pen returns to its original position it overshoots by approximately 1 per cent of the deflection but then returns to its position of rest without any further overshooting as shown in Fig. 3.5.

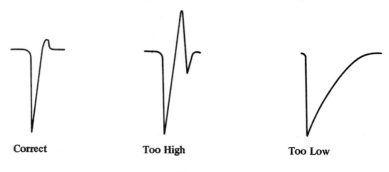

Correct   Too High   Too Low

Fig. 3.5

100 % control

This control is usually placed at the front of a spectrometer and operates an optical wedge or comb (Fig. 3.6) into or out of the sample beam. This is used to bring the recorder pen to a position of ≃ 90 per cent transmission at the clearest point in the sample's spectrum. Figure 3.7 shows the result of correct use of the 100% control whilst Fig. 3.8 shows the result of its incorrect use.

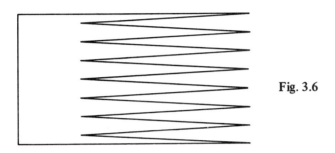

Fig. 3.6

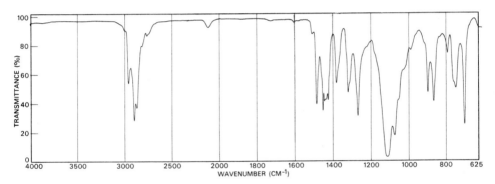

Fig. 3.7

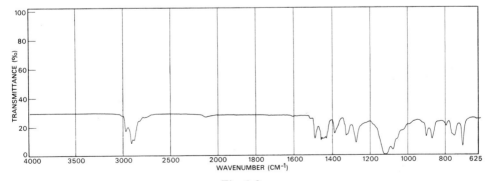

Fig. 3.8

54

Slit control

This control is used to adjust the resolving power of the spectrometer to give you the desired resolution in your spectrum. The need to adjust was described on page 44. (See Spectra 1 and 2).

This control has another important use. If you have to run a spectrum and very little energy is transmitted, such that attenuation of the reference beam is necessary, then because there is only a little energy reaching the detector the gain has to be turned up high. When this is done there may be excessive noise on the trace (see Spectrum 3). If the slits are now adjusted (made wider), more energy is allowed to pass to the detector and this allows you to turn the gain back down to a more normal value. This is shown in Spectrum 4. Note that you do not have to touch the attenuation of the spectrometer when you change the slit width.

We are now in a position to use a commercial double beam spectrometer. Switch on the instrument and allow a few minutes to warm up (see manufacturer's instructions for this). We now wish to run a spectrum of polystyrene film, which all manufacturers normally provide with their instruments. This film is used by manufacturers as it is a convenient method of providing a standard sample which will demonstrate the performance of the spectrometer and because its band positions are accurately known.

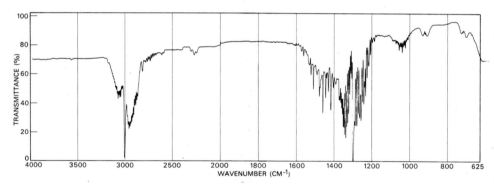

Spectrum 1

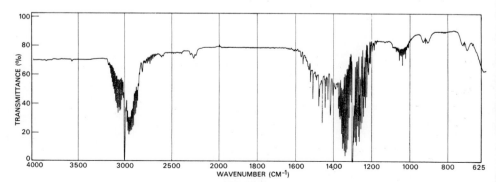

Spectrum 2

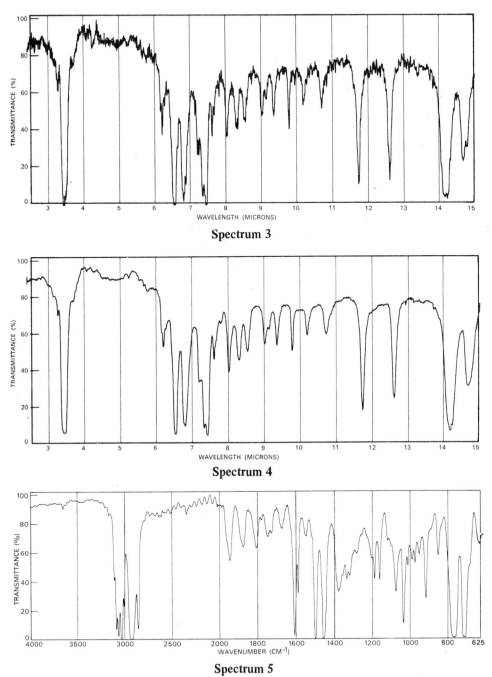

Spectrum 3

Spectrum 4

Spectrum 5

Question

To obtain a good spectrum such as that published by the manufacturer in his manual (see Spectrum 5), what procedures are necessary so that you may be sure of matching that spectrum exactly?

Please try to answer before you turn to → page 56.

The steps that you need to perform before you obtain a good standard spectrum are:

1. Check the attenuation 100 % control.

2. Check that the slit is on N (Normal).

3. Check, and if necessary adjust, the gain.

Part 4 starts on → page 57.

Part Four

BASIC THEORY

LET US see in what portion of the infrared spectrum you could expect atoms which are vibrating to absorb energy to give an absorption band.

Let us take as an example models of diatomic molecules (Fig. 4.1).

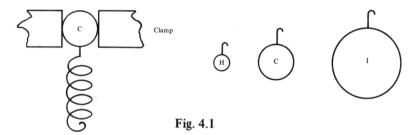

Fig. 4.1

The diatomic molecules under consideration are C–H, C–C and C–I. The carbon atom which is common to the molecules is held in a clamp and the bond, which is represented by the spring, is permanently attached to the carbon atom. The other atoms have hooks on them so that they may be attached to the spring. The mass of the weight to be attached is proportional to the mass of the atom. When the hydrogen is attached to the spring, stretched a little and released, it will vibrate at a particular frequency. This frequency is represented by v_1. If this mass is replaced by the carbon, then because it is a different mass the vibration of the spring will be different (slower) and the new frequency will be represented by v_2. If we now add the iodine atom, an even slower frequency of vibration will be observed v_3. The frequency of the vibration is inversely proportional to the masses vibrating. In the infrared spectrum the vibrations of these atoms will occur at

C–H, v_1 = 3000 cm^{-1} C–C, v_2 = 1000 cm^{-1} C–I, v_3 = 500 cm^{-1}

Question
If the diatomic molecule consisted of C–O, near to which of the three frequencies would you expect to find an absorption band?

1.   3000 cm^{-1} → **page 65**

2.   1000 cm^{-1} → **page 60**

3.   500 cm^{-1} → **page 62**

Your answer is incorrect. You will remember on page 60 we said that two dissimilar groups would vibrate at their normal diatomic frequency.

In the light of this information go back to → page 63 and choose the correct alternative.

An absorption occurring at 1600 cm^{-1} is most probably an absorption due to a $C = X$ group in an organic compound. Do you remember that on page 63 we said that if two similar bonds were joined together to make a triatomic unit the absorptions of this group would be displaced from the diatomic position? The absorption band due to the group $N = C = N$ is therefore not found here.

Return to → page 79, re-read the section and then choose another alternative.

Your answer is correct. The mass of the oxygen atom is very close to that of the carbon and so the frequency of vibration would be similar to that of carbon and be found near 1000 cm^{-1}.

Suppose now we connect two of these diatomic molecules having sufficiently different frequencies to make a hypothetical triatomic molecule e.g. C–C–H and C–C–I. We now have two bond 'springs' and there will be two ways to stretch each molecule.

$$\overset{\leftarrow}{H}-\vec{C}-C \qquad\qquad \overset{\leftarrow}{C}-\vec{C}-I$$

$$H-\overset{\leftarrow}{C}-\vec{C} \qquad\qquad C-\overset{\leftarrow}{C}-\vec{I}$$

$\leftarrow\rightarrow$: direction of motion of the atoms when the two 'springs' are vibrating.

The bonds in the H–C–C molecule will be found to vibrate practically independently i.e. when the C–H bond vibrates the C–C hardly changes, and the spectrum of this compound would show two absorption bands at 3000 cm^{-1} and 1000 cm^{-1}. Similarly in the C–C–I molecule the two vibrations are shown to be almost independent of each other and to occur very close to their diatomic positions. Because of this the spectrum would show two bands at 1000 cm^{-1} and 500 cm^{-1}.

Question
Where would you expect to find the absorption bands of this group of atoms: H–C–I?

1. 1000 cm^{-1} ; 500 cm^{-1} → page 68

2. 2000 cm^{-1} ; 500 cm^{-1} → page 72

3. 3000 cm^{-1} ; 500 cm^{-1} → page 63

Your answer that spectrum No. 2 contains a C=N group is incorrect. The spectrum is that of butyronitrile and contains a band at 2245 cm^{-1}. This band represents the C≡N. There is no band in the C=N region.

Return to → page 66 and select the correct alternative.

In the text we said that the frequency of vibration was inversely proportional to the mass vibrating. The mass of an oxygen atom is very much lighter than the iodine mass, but is more similar to another atom discussed in the text.

In the light of this fact go back to ← page 57 and choose a better alternative.

You are correct. The two frequencies would be 3000 cm⁻¹ and 500 cm⁻¹.
The two parts of the group of atoms would vibrate independently. There
would be an absorption due to the H–C portion at 3000 cm⁻¹ and an absorption at 500 cm⁻¹ due to the C–I. If we were to link two identical bonds together, say, to make a C–C–C molecule we would still get two absorption
bands but these would involve both of the bonds vibrating. The two frequencies would arise from an in-phase and an out-of-phase vibrational
motion.

$$\underset{\substack{\text{symmetric}\\\text{in-phase}}}{\overset{\leftarrow\quad\rightarrow}{C-C-C}}\qquad\underset{\substack{\text{asymmetric}\\\text{out-of-phase}}}{\overset{\leftarrow}{C-C-C}\rightarrow}$$

The two vibrations are usually termed symmetric and asymmetric respectively.
Because the two bonds do not move independently of each other we say there
is interaction between the groups and the frequencies of vibration will now
be displaced from the diatomic frequency.

Question
In which of these molecules will there be symmetric and asymmetric vibrations?

1.   H–C–I   → page 70

2.   C–C–I   ← page 58

3.   O–C–O   → page 66

An overtone occurs at a position in the spectrum equal to $n\nu$ where n is an integer of a fundamental and ν is the fundamental frequency. Your answer is correct for the first overtone absorption band observed at 250 cm^{-1}. However, this was not the question that was asked.

Go back to → page 71 and choose another alternative.

In the text we said that the frequency of vibration was inversely proportional to the mass vibrating. The heavier the mass the slower will be the frequency of vibration. Hydrogen, which is the lightest of all elements, absorbs at 3000 cm^{-1} when it is bonded to carbon. Which of the elements described in the text has a mass similar to the mass of the oxygen atom?

Think out the answer to this question then go back to ← page 57 and choose the correct alternative.

You are correct. The O—C—O vibrations will interact with each other to give symmetric and asymmetric motions and these will not occur at the diatomic position because this is now a triatomic vibration.

So far we have considered bonds of a similar type i.e. all single bonds. Now let us consider multiple bonds and their effect on the positions of vibration of the atoms.

The bonds which are found between carbon and nitrogen will serve as examples:

$$C-N \qquad C=N \qquad C\equiv N$$

If each bond is represented by a spring then the C—N will have one spring, the C=N will have two and the C≡N will have three. This means that the strength of the bond has been increased as you go from C—N to C≡N. If the strength of the springing has been increased and the atoms remain constant the frequency of vibration will also be increased. The frequencies of vibration of these groups are approximately:

C—N	C=N	C≡N
1070 cm^{-1}	1650 cm^{-1}	2250 cm^{-1}

Question
Look at the three spectra of nitrogen-containing compounds opposite. Which of these spectra is the one of the compound containing a C=N group?

1. Spectrum No. 1.   → page 73

2. Spectrum No. 2.   ← page 61

3. Spectrum No. 3.   → page 79

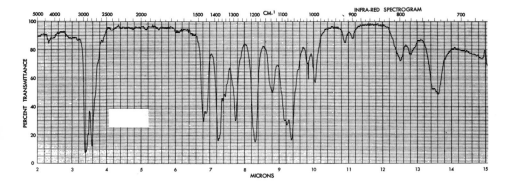

Spectrum No. 1

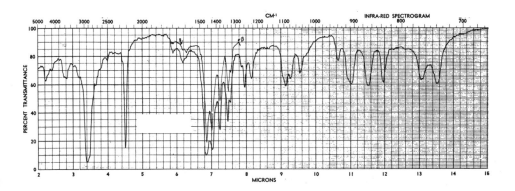

Spectrum No. 2

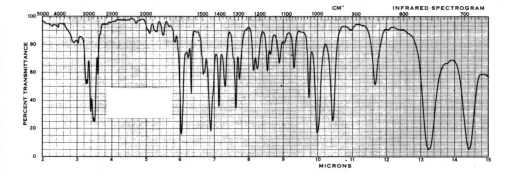

Spectrum No. 3

Spectra 1, 2 and 3 courtesy of Sadtler Research Laboratories. © Sadtler Research Laboratories, Inc (1972)

You have either been too hasty in your choice of an answer or you have not understood the text.

We described the triatomic molecules $H-C-C$ and $C-C-I$ in the text and said that the bonds would vibrate independently of one another. In the two examples the common position $(C-C)$ absorbs at 1000 cm^{-1} in both cases. In the question we asked about the absorptions you would observe for $H-C-I$. Don't you think that these would be the same as the non-common parts of the two molecules described in the text?

Return to ← page 60 and read it through carefully.

An absorption band occurring at 4000 cm^{-1} cannot have its first overtone at 2000 cm^{-1} because the first overtone is at twice the frequency of the fundamental and therefore a fundamental absorption at 4000 cm^{-1} would have its first overtone at 8000 cm^{-1}.

In the light of this information go back to → page 71 and choose the correct alternative.

You will notice that the compound is H—C—I and the groups are not the same on each bond. Because the groups are asymmetric there can be no inter-action and the absorption bands will be found in the normal diatomic posi-tion.

In the light of this information go back to ← page 63 and choose the correct alternative.

Correct. There will be six vibrations of the group i.e. asymmetric and symmetric stretches, in- and out-of-plane symmetric and asymmetric deformations.

The infrared spectrum has in its composition three types of absorption bands, namely, fundamental, combination and overtone.

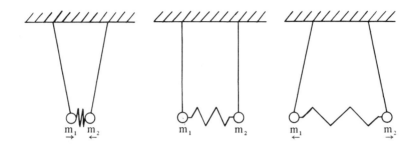

Fig. 4.2

Examples of these absorption bands may be derived if one considers the atoms in the molecule to be ball bearings with masses proportional to the mass of the atoms and connected by springs proportional to the bond strength. Consider the two masses shown in Fig. 4.2. If they are stretched apart and released it will be observed that they will vibrate at a certain natural frequency and this is the *fundamental* frequency for that system. The atoms are said to be vibrating with simple harmonic motion.

If a molecule with a fundamental vibration occurring at a frequency of v_1 is subjected to radiation at a frequency of $2v_1$, an absorption may also be observed in the spectrum at this frequency and this is called an *overtone* absorption band. This does not mean that the molecule itself is vibrating at $2v_1$, but the fundamental is being excited by the radiation at twice its frequency and such bands are generally much weaker than the fundamental concerned. If a molecule has two different fundamental absorptions at v_x and v_y it is possible that an absorption may be observed at frequencies corresponding to $v_x + v_y$ or $v_x - v_y$ and these are said to be *combination* bands. They will generally be weaker than the fundamentals concerned.

Question
What would be the frequency of the fundamental absorption if its first overtone absorption was observed at 2000 cm⁻¹?

1.	4000 cm⁻¹	← page 69
2.	1000 cm⁻¹	→ page 81
3.	500 cm⁻¹	← page 64

You have either been too hasty in your choice of an answer or you have not understood the text.

We described the triatomic molecules $H-C-C$ and $C-C-I$ in the text and said that the bonds would vibrate independently of one another. In the two examples the common position $(C-C)$ absorbs at 1000 cm⁻¹ in both cases. In the question we asked about the absorptions you would observe for $H-C-I$. Don't you think that these would be the same as the non-common parts of the two molecules described in the text?

Return to ← page 60 and read it through carefully.

Spectrum No. 1 is that of triethylamine. Please note that compounds containing C≡N are called cyanides or nitriles, compounds containing C=N are called imines and compounds containing C−N are called amines.

Return to ← page 66 and select the correct alternative.

You are correct. The N=C=N group gives the absorption band occurring at 2150 cm^{-1}.

Vibrations other than the stretching of a bond can and do occur. These are called deformations (δ) and they refer to the changing of the bond angle between the atoms of a molecule. These deformations occur at frequencies lower than those of the stretching vibrations.

There are some general descriptions of deformational vibrations which are applicable to $-(XY_2)-$ groups of atoms.

The terms used are scissors, rock, wag and twist. These four motions may be subdivided into in-plane and out-of-plane motions of the atoms. Let us consider them in order and draw some everyday analogies.

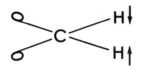
In-plane symmetric deformation
(scissors)

If you think of the C as the pivot and the H as the points of a pair of scissors then there is a plane through all of the atoms and the motion is in-plane.

In-plane asymmetric deformation
(rock)

Consider this as follows: the C is a person sitting in a rocking chair and the H as the tips of the rockers. If you consider this to be as drawn in the diagram then there is a plane through all of the atoms. This is a poor group frequency as all atoms move.

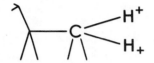
Out-of-plane symmetric deformation
(wag)

Consider the C to be the body of a dog and the H to be the tail. Then when the dog wags its tail the motion is from side to side and out of the plane of the dog (the plane of the paper in the diagram). This is also a poor group frequency.

Out-of-plane asymmetric deformation
(twist)

This direction of motion is best considered from a top view and is self explanatory.

Question
Look at the diagram of the CH_2 ClBr molecule and consider the CH_2 part only. What is the **total number** of vibrations which may occur?

1.   4   → page 82
2.   5   → page 80
3.   6   ← page 71

Consider **1**.

The arrows give the direction of motion of the C atoms so let us now move them to the positions indicated.

Although there is now a greater distance between the C atoms the symmetry of the molecule is the same, therefore it is infrared inactive as there is no change in dipole moment.

Consider **2**.

The arrows give the direction of motion of the H atoms so let us move them to the positions indicated.

The molecule now looks more distorted. However, it is still the same on both sides of the centre of the C=C bond; the dipole moment has not changed, it has the same symmetry and is infrared inactive.

Consider **3**.

Move the H atoms and you now get

Now there is a difference across the molecule. The right hand side has the H and Cl close together, the left hand side has the H and Cl wide apart. There is a change in dipole moment across the molecule and the vibration is infrared active.

Part 5 starts on → page 83.

An absorption occurring at 3000 cm^{-1} is unlikely to be the required absorption band as we said that C—H groups absorbed at this position. You were correct in thinking that displacement from the diatomic position would occur, but it would not be as large as you suggest.

Go back to → page 79 and choose another alternative.

The answer to the question on page 81 is : 1 and 2. Did you get the correct answers? If you did not, turn to ← **page 76** to see how they were worked out. If you got both correct and are sure you understand why they are inactive continue the programme by turning to **Part 5** which starts on → **page 83**.

You are correct. The compound is benzylidenimine N-methyl. It contains a
$C=N$ group and will have an absorption band at approximately 1650 cm^{-1}.
The other two spectra did not have this absorption band. Did you notice the
absorption bands characteristic of the $C-N$ and $C\equiv N$ in the other two spectra
at 1070 cm^{-1} and 2245 cm^{-1} respectively? The compounds were (1) triethylamine, and (2) butyronitrile.

If we combine a bond with another of sufficiently different frequency the
bonds will vibrate separately e.g. $N=C-N$ and $N\equiv C-N$ will show two absorption bands near to 1650 and 1070 cm^{-1} and 2250 and 1070 cm^{-1} respectively.

If we now make a triatomic molecule of the type $N=C=N$ we can get symmetric and asymmetric vibrations of the molecule. This interaction will dis-
place the frequency from the double bond region and it will, in fact, move it
to a position very close to the $N\equiv C$ region. This means that we have to consider this type of group as a whole entity rather than two separate $C=N$
groups.

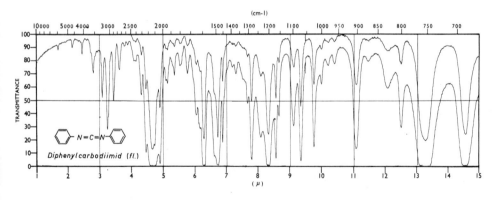

Spectrum 4

Taken from R. Mecke and F. Langenbucher, *Infrared Spectra of Selected Chemical Compounds*,
Heyden & Son, London, 1965.

Question
Spectrum 4 is of a compound containing a $N=C=N$ group; it is the spectrum
of N,N'-diphenylcarbodiimide. What is the frequency of the absorption band
which is characteristic of the group of atoms $N=C=N$?

1. 3000 cm^{-1} ← page 77

2. 1600 cm^{-1} ← page 59

3. 2150 cm^{-1} ← page 74

You have made a guess at the answer but you forgot that two similar bonds linked together give rise to two vibrations and so besides the four deformations there are two stretching vibrations — a symmetric and an asymmetric.

In the light of this information go back to ← page 75 and choose the correct alternative.

Good. The fundamental absorption would be at 1000 cm⁻¹.

An infrared absorption arises when there is a change in dipole moment caused by the vibrating atoms. The atoms of a molecule carry a small positive or negative charge. The positions of these charges vary relative to one another as the atoms vibrate. The *dipole moment* is the vector sum of these charges. If there is a net difference between the dipole moments of the vibrating atoms when they are at the extremities of vibration, the vibration is said to cause a change in dipole moment and hence give rise to an infrared absorption.

For example, CO_2 can vibrate as follows:

$$\overset{\leftarrow}{O}=C=\vec{O} \qquad \vec{O}=\overset{\leftarrow}{C}=\vec{O} \qquad \overset{\uparrow}{O}=\underset{\downarrow}{C}=\overset{\uparrow}{O} \qquad \overset{+}{O}=\overset{-}{C}=\overset{+}{O}$$

symmetric   asymmetric   O=C=O deformation
 stretch   stretch   667 cm⁻¹
              2350 cm⁻¹

These four vibrations are the four fundamental vibrations of CO_2. For CO_2 the C=O symmetric stretch does not give rise to a change in dipole moment, consequently radiation is not absorbed and no band is observed in the infrared spectrum for this vibration. The C=O asymmetric stretch causes a change in dipole moment, and so is observed. The bending of O=C=O bonds are identical motions but they occur in perpendicular planes. Such vibrations are called *degenerate* and appear in the same position in the spectrum. The two bands, due to the bend and the asymmetric stretch, constitute the fundamental spectrum of CO_2.

Question
Which of the following three vibrations are infrared **inactive**? (Arrows indicate the direction of the moving atoms; consider all others to remain stationary).

1.

2.

3.

To check your answers turn to ← page 78.

You have only counted the vibrations described in the text which are the deformations. It was stated on page 66 that other vibrations do occur.

Go back to ← page 75 and choose another alternative.

Part Five

SAMPLE PREPARATION

THE CONSTITUENT atoms of a molecule are constantly oscillating about their mean positions at clearly defined frequencies. When such an oscillation disturbs the dipole equilibrium of the molecule, radiant energy will be absorbed if the frequency corresponds to that of the oscillation. The region of the electromagnetic spectrum in which these molecular vibrations occur is called the infrared region.

Organic and some inorganic molecules can cause absorptions at a large number of frequencies; these compounds are said to have a complex infrared absorption spectrum. If the absorption bands of substances giving complex spectra are examined in the 4000 cm^{-1} to 650 cm^{-1} region it will be noticed that each compound has a different absorption spectrum. This region 4000 cm^{-1} to 650 cm^{-1} is called the fundamental or fingerprint region of the infrared. Taking an infrared spectrum of a compound in this region is like taking a person's fingerprint in that each person has a different fingerprint so each compound has a different infrared spectrum. Identification by an infrared spectrum is therefore a matter of sorting and matching the unknown spectrum with that of a known spectrum.

An infrared spectrum gives the chemist a permanent step-by-step record of his work. When carrying out a complex synthesis he can obtain a record of the changes taking place and a check on the purity of his reagents. There is also no fear of using up his sample by taking an infrared spectrum since infrared analysis is very sensitive and non-destructive.

Question
Which portion of the electromagnetic spectrum is used to obtain a fingerprint of an organic compound?

1. $2.5-15 \, \mu\text{m}$ → page 92

2. $0.1-6 \, \mu\text{m}$ → page 90

3. $0.1-100 \, \mu\text{m}$ → page 88

Only one of the windows which you have picked is in the list of six common windows.

Intrans 5 (magnesium oxide) does not cover the useful range to 15 μm which is required for most organic liquids. It is also expensive and would only be used in special work involving water or acids as solvents. Your answer is therefore wrong.

Go back to → page 93 and choose another alternative.

You are wrong. The shape of the cell has nothing to do with the question of why the cells for gases need to be longer than cells for liquids.

Re-read → page 97 and then attempt the question again.

You are correct. NaCl and KBr are the two most common window materials.

Liquids and gases need to be held between two windows so that the infrared radiations will be absorbed only by the sample and not the holder. The apparatus which is used for holding liquids and gases in fixed volumes is called a cell.

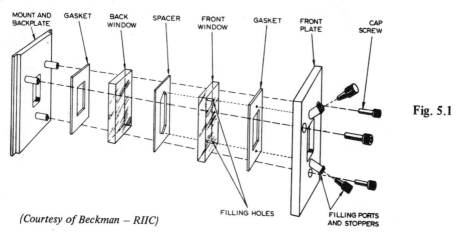

(Courtesy of Beckman – RIIC)

Fig. 5.1

Liquid cells consist of two windows sandwiching a thin film of liquid, sometimes purely a capillary film. Thicker layers require a spacing washer of an opaque inert material to separate the plates. The spacing washer has an aperture to allow the light to penetrate. The liquid is contained in the aperture space between the windows, both of which are sealed to the spacer to prevent escape of the liquid. One of the windows has two holes drilled through it to provide inlet and outlet, allowing the sample to be conveniently placed in the cell from a hypodermic syringe or capillary dropper. There are two types of liquid cells in use, demountable (Fig. 5.1) and permanent (Fig. 5.2). You need to know when to use each type and their properties are compared in the table opposite.

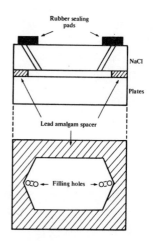

Fig. 5.2

Continue by answering the question on → page 87.

Permanent

Spacer may be made from lead amalgam or nylon. The seal for lead amalgam is made in the cold by the reaction of the amalgam with the halide. The seal for nylon is made by melting the nylon between the halide plates, pressing them together and then allowing the sandwich to cool.
Advantages. There is no fear of cracking the plates by over-tightening the holder. They may be used for very thick cells.
Disadvantages. Lead amalgam spacers cannot be used for cells $<50 \mu$m thick because of the difficulty of handling the spacer which is easily dissolved by the mercury.
The thickness of the cell is not known until it is made because both methods depend on pressing a soft spacer between the windows.
The cells are difficult to keep clean and therefore only have a short useful life.
The nylon spacers are used for thicknesses $<250 \mu$m.
The cells vary in thickness during their useful life due to etching of the windows by wet solvents.

Demountable

The spacer is usually made from Fluon. The seal is made by squeezing the plates together in a special holder. Because the Fluon is soft it can be pressed against the window to make a liquid-tight seal. The cell is always in a special holder.
Advantages. The cells are easy to clean by demounting, polishing and remaking the cell.
The thickness of the cell is easily and quickly changed without destruction of any portion.
Disadvantages. The plates may be broken by over-tightening the holder and care must be exercised when performing this operation.
The cell may leak especially if low boiling ($<40°$C) liquids are being examined.

Question

What type of cell is the most difficult to keep clean and is most likely to vary in thickness during use?

1. permanent → **page 97**

2. demountable → **page 100**

3. don't know → **page 89**

back ref. page 83

One of the wavelengths given (0.1−100 μm) falls outside the infrared region of the electromagnetic spectrum. The region covered is not found on the standard spectrometers. I also think that you have forgotten how to convert wavenumbers into wavelength.

Re-read ← page 11 and then turn to ← page 83 and work out your answer again.

This is not an easy question and can only be answered if the text has been read carefully.

Consider the essential difference between permanent and demountable infra-red cells. The windows of a permanent cell are stuck to the spacer and the cell cannot be taken apart without the windows breaking. Demountable cells can be taken apart if contaminated and rebuilt to their original conditions with the minimum of bother.

Now return to ← page 86 and before choosing another answer re-read the section in the light of the information above.

One of the wavelengths given (0.1−6 μm) falls outside the infrared region of the electromagnetic spectrum. The region covered is not found on the standard spectrometers. I also think that you have forgotten how to convert wavenumbers into wavelength.

Re-read ← page 11 and then turn to ← page 83 and work out your answer again.

It is easier to spread a liquid than a powder over a window but this is not the real reason why a liquid is used when examining a solid sample. One would not go to the trouble of grinding the sample if it was just to add a binding agent before examining the sample.

In view of these conclusions, go back to → page 99 and choose a better alternative.

You are correct. The wavelength range is 2.5 μm to 15 μm for the fingerprint of an organic compound.

Gases and liquids must be contained in cells with end plates which are transparent to the radiation range to be observed. There are a number of suitable materials, but only a few are in widespread use. The limitation on the uses of compounds as windows are their transmission range, solubility in water, softness and the cost of the windows.

In spite of their individual limitations there are sufficient window materials available to cover the full range of the infrared region. The window materials are suitable for any type of compound one wishes to examine. The table opposite includes most of the common window materials. Long wavelength limits are marked on the table.

These eleven window materials will cover most applications in the infrared. Of these, six are frequently used in infrared laboratories. The six are: glass, sodium chloride, potassium bromide, silver chloride, caesium iodide and polyethylene.

Question
Search the table carefully and compare the properties of the following pairs of windows. Which do you think would be in common use in infrared laboratories?

1.   k,i,   → page 95

2.   b,e,   ← page 84

3.   e,g,   ← page 86

	0.5 μm 10 μm	100 μm
a	3.5 μm Glass	Readily available in suitable plates but limited in useful range.
b	8.5 μm MgO Intrans 5	Hard, insoluble, obtained as a compact. Expensive.
c	14 μm Zinc sulphide ZnS Intrans 2	Hard, insoluble, obtained as a sintered compact. Used for water solution work. Expensive.
d	15 μm Silicon Si	Insoluble in water and acids. Resists thermal shock and the transmission is not affected by temperature.
e	18 μm Sodium chloride NaCl	Cheapest and most common window material. It is easily scratched and cleaves easily.
f	28 μm Silver chloride AgCl	Soft, ductable and may be rolled into thin sheets. It reacts with some metals but is not hygroscopic. It is insoluble but darkens slightly on exposure to visible and ultraviolet light.
g	30 μm Potassium bromide KBr	Readily available as polished plates, hygroscopic, cleaves and scratches. A common window material.
h	40 μm Thalium bromoiodide KR.S5	Moderately expensive, very slightly soluble in water. It is soft and is used as an ATR* crystal material. Does not cleave and is toxic.
i	20 μm Caesium 60 μm iodide CsI	Does not cleave, moderately expensive. Resists thermal shock, hygroscopic, soft and scratches easily. Used frequently in the 20–60 μm region.
j	20 μm Polyethylene	Cheap, readily available, used for cold solvents and mulls for all substances at wavelengths >20 μm.
k	diamond	Hard, expensive, used as a window material in specialised applications and as far infrared detector windows.

* see Appendix A.

A multi-reflection cell is a device for obtaining a strong spectrum from a small quantity of gas enclosed in a relatively small volume cell.

The light is reflected backwards and forwards inside the cell many times before it emerges and hence has 'seen' much more of the sample than would have been the case had it just passed directly through the cell.

In the light of this information go back to → page 102 and choose the correct alternative.

As described in the table, diamond windows are only used in specialised applications and are not in common use.

The choice of caesium iodide is also poor because this material is usually used in the range 20 to 60 μm which is outside the range of most of the infrared instruments and is mainly used by metal-organic and inorganic chemists. This choice of windows is wrong.

Return to ← page 93, re-read the table and then attempt the question again.

Whilst it is true that CS_2 and CCl_4 dissolve many organic substances, this fact is also true for many other common organic solvents and thus does not set CS_2 and CCl_4 apart.

Let us reconsider the part a solvent must play in the infrared.

The substance we are interested in must be soluble in the solvent as you suggested but furthermore it must not mask the sample spectrum more than is necessary. Whilst it is impossible to obtain a solvent which is completely devoid of absorption bands in the infrared, some solvents do approach this ideal.

In the light of this information go back to → page 105 and choose the correct alternative.

You are correct. A permanent cell is more difficult to clean and is most likely to vary in thickness during its lifetime. Once it is made and in use the thickness of the cell is determined by the amount of wet solvents you use in the cell. It cannot be taken apart, the windows cleaned and a new spacer placed in the cell as easily as a semi-permanent cell.

Let us now look at another type of sample which is contained in a cell, namely a gas sample.

There are fewer molecules in a gas for any given volume than there are in a liquid of the same volume. It is the number of molecules absorbing the infrared radiation which decides the optimum size of a cell i.e. it is the density of the sample which governs the cell size. A 10 cm thick cell is the most common size for gas samples since it gives a reasonable absorption level for the majority of gases and vapours with pressures in the cell equal to or less than one atmosphere.

Because the gas cells have to be so long and the optics of the instrument are such that the light beam is convergent, a cell which is shaped to the beam will have a minimum volume and still let through a maximum amount of light.

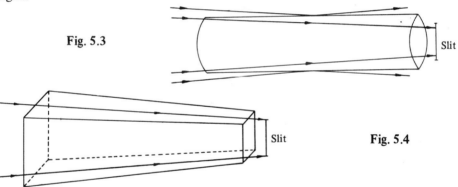

Fig. 5.3   Slit

Slit   Fig. 5.4

Figures 5.3 and 5.4 show the paths of the light in an instrument sample compartment. You can see that with cell 5.4, which has the same volume as cell 5.3, only about half of the light from the source is being used. The other half is wasted because the light never reaches the entrance slit of the instrument. The fact that only half of the light is used means that the reference beam of the spectrometer has to be similarly compensated. This results in the energy at the detector being smaller and consequently there is more electronic amplifier noise on the trace.

Question
Why do cells for gases need to be so much longer than cells for liquids?

1. Because the gas absorption bands are weak.   → **page 104**

2. Because the density of a gas is so much less than
    that of a liquid.   → **page 102**

3. Because minimum volume gas cells are used   ← **page 85**

H

98

Let us consider the arrangement in the spectrometer. A small volume of an organic gas has been collected in a container, sealed, and then infrared radiation passed through the container. Now all that can happen to the gas is that it warms up slightly; this will increase the partial pressure of the gas in the container but not the amount. Your answer is therefore incorrect.

In the light of this information go back to → page 102 and choose the correct alternative.

Correct. CS_2 and CCl_4 have relatively few bands in the fingerprint region.

Let us now look at the ways of examining organic solids (usually crystalline materials). If you cannot run solution spectra because of interference from the solvent absorptions then you must try to examine these compounds as solids and this is where new problems of sample handling occur.

If you take a crystalline material and grind it, you will produce a fine powder, but infrared radiations will still find it difficult to pass through this powder. If it is placed on a window the radiation will be scattered in all directions from its surfaces; this will cause insufficient radiation to enter the sample or spectrometer to give a spectrum. To overcome this scattering one must try to eliminate this 'particle' sample by adding to the powder a substance which will not scatter the radiation. The addition of a liquid to a powdered solid is a way of reducing the refractive index difference between the solid and its surroundings and hence cut down the scattering of the radiation. If the finely ground solid is covered with a liquid of a similar refractive index the incident radiation will not be scattered, as it will 'see' a homogeneous sample and not a series of discrete particles. When the radiation is scattered from the surface of a solid none will penetrate and no spectrum will be obtained.

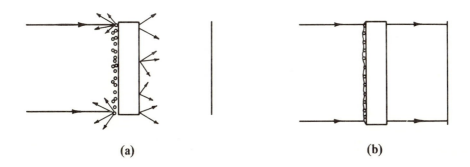

(a)   (b)

Fig. 5.6   Diagram of incident radiations on a powder. (a) Powder on a window. (b) Powder dispersed in liquid on a window.

Question
Why are powdered crystalline samples dispersed in a liquid and not placed directly into the infrared spectrometer beam?

1.   Because this is a method of reducing the
     refractive index of the sample.   → page 103

2.   Because this is a method of reducing scattering
     of the radiation such that more radiation goes
     into the sample and the spectrometer.   → page 108

3.   Because it is easier to spread liquid over a plate
     (window) than it is a solid.   ← page 91

You are wrong. A demountable cell can be taken apart and the windows cleaned. The fluon spacer can easily be replaced if damaged or if it has worn thin.

In the light of this information go back to ← page 86 and choose the correct alternative.

Most organic solvents are mobile and would run easily into an infrared cell so this fact does not separate CS_2 and CCl_4 from a host of other solvents.

A good infrared solvent must be one which will interfere least with the spectrum of the sample. Ideally it should have no absorption bands in the infrared. Whilst this is not completely possible, some solvents do, in fact, come close to this ideal.

In the light of this information go back to → page 105 and choose the correct alternative.

Correct. It is the density of a compound which determines the type of cell you would use.

Special purpose gas cells. When only a small amount of gas is available don't you think it would be a good idea to use the molecules of a gas over and over again so bringing the absorptions up to liquid type intensities? A specially constructed cell is made for this purpose.

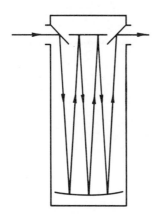

Fig. 5.5

The short distance of a gas cell may be effectively lengthened by passing the light backwards and forwards through the same volume of gas a number of times before admitting the light to the instrument. This arrangement of mirrors (Fig. 5.5) has effectively lengthened the path of the gas cell by six times and if the two mirrors at the entrance and exit are adjustable then the lengthening of the cell is also adjustable. Path lengths of up to 10 m are easily obtained by this means. This allows very small amounts of a compound to be identified. Note that the diatomic gases O_2 and N_2 do not have any absorptions in the infrared and may be used as diluents so that the cell may be at atmospheric pressure.

There are a few points which need careful attention when you are using these cells. Some light can be lost at each reflection and scattering of the light will occur if the mirrors are dirty. This means that although these cells will increase the effective path length they also lose a little of the useful energy and so need careful positioning in the spectrometer.

Question
By passing the light through the gas a number of times as in a multi-reflection gas cell do you

1.   increase the effective path length of the cell?   → page 105

2.   increase the concentration of the sample in the cell?   ← page 98

3.   Don't know.   ← page 94

The refractive index of a sample is a physical property of the sample and cannot be changed by the addition of another substance. All that you do is *reduce its relative refractive index,* but one is not measuring the refractive index of a sample by infrared spectroscopy.

In view of these conclusions, go back to ← page 99 and choose a better alternative.

The size of the absorption band is a function of a number of variables. The answer you gave is not the whole truth and a much better reason is given in the text.

In the light of this fact go back to ← page 97 and choose a better alternative.

Correct. One uses a multi-reflection gas cell to increase the effective path length of the cell; it is used when only a small gas sample is available.

An infrared spectrum may be obtained from any type of compound i.e. solid, liquid or gas. The ways of handling the samples are different because of the physical state of the substance and it is the handling of samples which we will now discuss.

Most pure organic liquids have extinction coefficients in the infrared region such that a sample thickness of between 10 μm and 100 μm has an absorption spectrum of suitable intensity for identification purposes. Liquid cells for use in this region therefore usually have a cell thickness of between 10 μm and 1000 μm. The portion most commonly used for identification purposes is the 4000 to 600 cm^{-1} and of the solvents, CCl_4 and CS_2 are most useful in that they have very few absorption bands. Where it is necessary to run solution spectra between 4000 and 600 cm^{-1}, the portion 4000 to 900 cm^{-1} may be run in CCl_4 and 900 to 600 cm^{-1} in CS_2 with little interference in the spectrum of the compound by the solvents used. This assumes that the compound is soluble in both CCl_4 and CS_2 and does not react with them.

Question
Why are CS_2 and CCl_4 very good solvents for use in infrared spectrometry?

1.   Many organic substances will dissolve in them.   ← page 96

2.   They are very mobile and easily run into the   ← page 101
     narrow cells.

3.   They have very few absorption bands and so   ← page 99
     interfere least with the solute spectrum.

back ref. page 118

I am sorry but your answer is not correct. Just mixing finely ground sample and KBr and pressing into a disc is not the complete answer. KBr is hygroscopic and the sample may be wet. Care has to be taken to ensure that only dry materials are used. The correct answer is to mix only dry sample and dry KBr and press into a disc.

In the light of the statements above go back to → page 118 and select the correct alternative.

Here is a table showing the methods of preparation and the types of material for which they are useful.

Method of preparation	Sample type
Casting: dissolving the polymer, casting a film onto a plate and removing the solvent.	Soluble polymers: polystyrene and polymethyl methacrylate
Hot pressing: heating the polymer until it softens then pressing in a hydraulic press.	Polymers which soften on heating (polythene).
Microtoming: slicing a thin film with a blade.	Cross-linked rubbers.
Melt: allowing the sample to melt by placing it between two hot salt plates.	Many low melting point solids.

Now turn to → page 121 and continue the programme.

Correct. It is to reduce the scattering of the radiation that a liquid dispersant is used. The addition of a liquid oil to a solid sample makes the solid 'transparent' to the radiation. The radiation has to penetrate the solid and go into the spectrometer before absorptions of the solid may be measured and a spectrum obtained.

Method. Approximately 50 mg of the organic solid sample is finely ground using a mortar and pestle. One drop of the liquid dispersant is now added and the grinding continued; further drops of the dispersant may be added as necessary to obtain a cream-like consistency. The *mull* has now been prepared, but because it is an oil it must have some support to enable a spectrum to be run. The mull is scraped out of the mortar, placed onto a KBr window and another window is placed onto the mull. This squashes the mull and spreads it across the windows making a thin sandwich with the mull as the filling.

This is the easiest way of preparing a sample so that you may obtain its infrared spectrum. As it is a widely used technique it is important that you learn where the dispersant's bands come so that you are not misled by them when you are matching your spectra.

The liquid media used are nujol (liquid paraffin), hexachlorobutadiene and Kel F (Spectra 1, 2 and 3 opposite). The reason that there are a number of liquid dispersing agents is that they all have absorption bands in the infrared and you choose the one which will interfere least in the spectral region you are examining.

Nujol (liquid paraffin) is a saturated aliphatic hydrocarbon and contains CH_2 and CH_3 groups only. When this is used as a dispersant for an organic compound it will have absorption bands near 3000 cm^{-1} for the CH stretch, 1400 cm^{-1} for the CH deformation and 720 cm^{-1} for the CH_2 rock. If the sample you are examining contains any CH_2 or CH_3 groups you can see that they will not be observed because the nujol which is present in a mull in the highest concentration will mask these absorption bands.

Hexachlorobutadiene (C_4Cl_6) and Kel F contain no hydrogen atoms. Kel F is a halogenated oil which is less volatile and not as toxic as hexachlorobutadiene. These oils do not contain hydrogens and so do not absorb in the regions where CH absorptions occur. They contain C–Cl groups and so they have absorptions where the C–Cl absorptions occur near to 700 cm^{-1}. Although they do not interfere with the CH absorptions they will interfere with other absorptions of a compound. This interference (masking) of absorption bands makes the use of two different dispersants necessary if a full spectrum of the compound is required.

Continue by answering the question on → page 109.

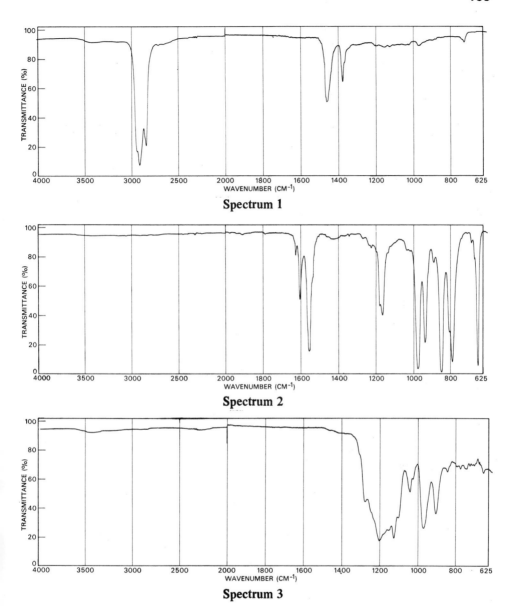

Spectrum 1

Spectrum 2

Spectrum 3

Question
A mull is the dispersion of a finely ground solid in a liquid (oil). If you wished to obtain a complete spectrum of the solid 1,2-diphenylethane, would you

1. use nujol as a liquid dispersant and obtain a good spectrum? → page 113

2. use Kel F as a liquid dispersant and obtain a good spectrum? → page 120

3. run two spectra — one of the sample dispersed in nujol and one of the sample dispersed in Kel F? → page 118

If the compound is *inter*molecularly hydrogen-bonded then on dilution with a non-polar solvent the separation of the solute molecules is varied and the hydrogen bonds holding the solute together are broken. Because some bonds are broken before others there will be molecules in different states of association, some hydrogen-bonded and others non-bonded.

In the light of this information go back to → page 121 and choose the correct alternative.

I am sorry but your answer is not correct. When trying to take a spectrum of a solid it is necessary to ensure that the sample size is small so that it does not scatter light and that the sample is homogeneously mixed with the disc material. This can only be achieved by using powders and mixing them by good grinding. Furthermore, a few large lumps of sample spread throughout the disc would allow large amounts of light to enter the spectrometer without passing through the sample.

In the light of this information go back to → page 118 and choose the correct alternative.

Well done; of the alternatives given only with a substituted benzene ring compound could you be sure that no interaction between the halide and the sample would take place.

We have described some methods for dealing with common solid, liquid or gaseous samples. We will now look at some other methods for more difficult samples. There are three main ways of dealing with polymer samples.

Casting. The polymer is dissolved in a solvent and then cast as a thin film on to a salt plate or microscope slide. After drying in an oven the sample can be examined either by removal from the microscope slide or directly on the salt plate. Typical polymers which can be treated in this way are polymethyl methacrylate and polystyrene.

Pressing. If the polymer can be softened without decomposition a film can be produced by heating the polymer to its softening point and pressing in a hydraulic press until a film thin enough for use is produced. A good example of this type of polymer is polyethylene. (The majority of thermoplastic polymers can be dealt with in this way).

Microtoming. Some polymers are soft enough to enable a thin film to be sliced from them using a microtome. Microtome sections can be taken from most soft polymers. Typical samples which may be examined in this way are cross-linked rubbers.

Another method which should perhaps be mentioned here is applicable to any low melting point solid (50°C). The solid is placed between two salt plates and heated until it melts. On cooling the solid sample may be examined. This is said to be running the sample as a 'melt', when in fact it is usually solid, the term 'melt' describing the method of preparation. Examples of samples examined in this way are phenol and waxes.

Question
1.   What type of sample would you examine as a microtomed section?

2.   To what does the term 'melt' refer?

3.   How many ways are there of preparing films of polymers?

4.   Which samples are examined as 'melts'?

5.   What method of sample preparation can be used for thermoplastics?

Write down your answers. On completion, turn to → page 119 and check them.

You have chosen the wrong answer. It was stated in the text that nujol is an aliphatic hydrocarbon and will therefore mask all C–H bands of an organic sample. Hence some other mulling agent containing no C–H groups must be used if you are to see these bands.

Go back to ← **page 109** and choose another alternative.

I

back ref. page 118

It is desirable when making a disc that all of the components are dry; you are therefore correct.

The method of preparing a halide disc is as follows:

1. 2 mg of the sample are finely ground using a mortar and pestle.

2. 200 mg of the dry halide are added to the sample in three equal portions and ground rapidly and well after each addition so that a fine powder, with the sample evenly dispersed in the halide, is obtained.

3. The powder is now placed in a special die. This is assembled, then connected to a vacuum pump and evacuated for 1 to 5 minutes.

4. The evacuated die is now placed in a press and the pressure on the die is increased until the halide sinters, (usually 10 to 15 tons pressure on the ram). The sintered mass is called a disc.

5. The vacuum is released and then after another half-minute the pressure on the die is released and the disc is removed from the die.

6. The disc should be handled as little as possible and quickly placed into a special holder ready to be fitted into the spectrometer. In part 4 above we mentioned a press. This is usually large, heavy and what is more important, expensive. A cheaper method for sintering the disc is detailed below (Fig. 5.7).

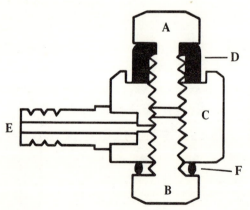

Fig. 5.7   A and B are the two bolts, C is the cylinder, D is the rubber seal, E is the evacuation tube and F is the rubber O-ring.

(Courtesy of Beckman – RIIC)

The barrel has one bolt threaded into it (bolt A) and then one-third of the halide mix (see points 1 and 2 above) is poured into the other end. Bolt B is now screwed into the barrel as tightly as possible. This should be sufficient to sinter the halide. The bolts are withdrawn and the barrel acts as the disc holder and is placed in the instrument ready for the spectrum to be obtained. The disc is removed by washing out with water. The barrel is then dried ready for the next sample.

Because KBr and NaCl are ionic solids and hygroscopic there are a few advantages and disadvantages to be remembered when using this technique.

Advantages

The KBr and NaCl have no absorption bands in the fingerprint region.

A sample in the form of a disc can be transferred from place to place and may be stored for reference purposes.

The refractive indices of the salts used are similar to those of a large number of solids. This reduces the amount of scattered light and reflection losses incurred when the refractive indices are dissimilar.

The disc method requires less sample than the mull method.

Disadvantages

The salts are hygroscopic and will pick up water from the atmosphere. This can be a disadvantage when studying N—H or O—H absorptions.

A poor spectrum will be obtained if both sample and halide are not properly ground and thoroughly mixed. The spectrum will appear as though not enough sample has been used and only the strongest absorption bands will show.

Cloudy, opaque discs will be obtained if the sample and halide are not dried or sintered correctly. This causes scattering of the infrared radiation at the short wavelength end of the spectrum.

The halides are ionic and so may react with ionic samples.

High pressures are used to form the discs and in exceptional circumstances decomposition or rearrangements of the sample may occur.

The choice of whether to use a disc or a mull for a particular sample is generally a personal one. It is sometimes determined for you by the reactivity of the sample or the amount of sample available. **Whichever technique one adopts a sample which has been finely ground and thoroughly mixed gives the best result.**

Question

If you are given the three materials below to examine as a KBr disc, from which one would you expect the least trouble?

1. Phenol → **page 122**

2. 1,2-diphenylethane ← **page 112**

3. Aniline hydrochloride → **page 117**

You are wrong. If a hydrogen bond can be broken then you would expect to see both the bonded and non-bonded hydrogens. This would give two absorption bands, not one.

Re-read → page 121 again and then select the correct alternative.

I am sorry but you are wrong. The halide salts used are ionic themselves and may react with other ionic materials.

Halide discs are notoriously reactive and care must be taken in choosing suitable samples for them.

Go back to ← page 115 and choose another alternative.

Well done, you have understood that a nujol mull alone is not sufficient to obtain a spectrum of every band of a compound.

There is a second, widely-used, method for the preparation of solid samples. When some of the common window materials e.g. NaCl, KBr or CsI are finely ground and then subjected to pressure they will sinter into clear homogeneous discs. If a small amount of a finely ground sample is introduced into the finely ground window material and the mixture pressed, the resulting disc will be suitable for infrared examination.

Unfortunately this technique has two major drawbacks: the necessary dies and presses are often expensive and the halides of sodium and potassium are hygroscopic and the discs may become opaque because of water pick-up.

Question
When you are examining a sample as a disc do you:

1. mix the sample with KBr and press into a disc?   ← page 111

2. mix the finely ground sample with ground KBr,
    and then press into a disc?   ← page 106

3. mix the finely ground, dry sample with dry, ground KBr
    and then press into a disc?   ← page 114

Answers
1. Cross-linked rubbers, soft polymers.

2. Method of preparation of low melting solids.

3. Three (a) casting from solvent
             (b) microtoming
             (c) hot pressing

4. Low melting solids (waxes).

5. A film may be hot pressed.

If your answers are correct please turn to → page 121.
If any of your answers were wrong and you would like to think about them again turn to ← page 107.

Your answer is wrong. The liquid Kel F has absorption bands in the finger-print region that will probably mask some of the sample's absorption bands.

Go back to ← page 109 and choose another alternative.

Now that we have dealt with the sampling techniques, let us study a type of sample which we will encounter and which may present some difficulty, namely polar compounds.

Polar compounds such as alcohols, amides and carboxylic acids are difficult compounds to run as solids or solutions. There are two main reasons why they are difficult.

(a) The solids are occasionally very crystalline with high refractive indices. This makes grinding and matching of the refractive indices difficult (see later for faults on grinding).

(b) Their solution spectra are difficult to run because these compounds are not very soluble in non-polar compounds. This means that the solvent spectrum will have absorption bands which may mask the solute spectrum because of their similarity. If a non-polar solvent can be used then it may cause the solute to show more absorption bands than those shown by the solid spectrum. This change in the spectrum is due to the fact that the compound is normally *inter*molecularly hydrogen-bonded and on going into solution this hydrogen bond may be broken. This causes changes in the spectrum, especially in the O–H or N–H stretching region. (See Fig. 5.8 for the changes which occur when an alcohol is dissolved in CCl_4 – spectra vary with the dilution). *Intra*molecular hydrogen bonds, however, are not affected by dilution and consequently the spectrum does not change. This fact enables one to differentiate between *intra-* and *inter*molecularly hydrogen-bonded compounds.

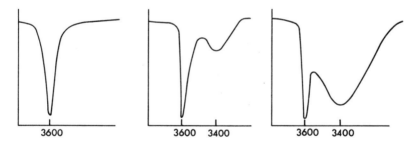

Fig. 5.8   Increasing concentration of alcohol in CCl_4.

Question
Would you expect there to be a change in the absorption bands seen in a non-polar solution spectrum of an *inter*molecularly hydrogen-bonded compound when the concentration is varied?

1.   Yes   → page 123

2.   No   ← page 116

3.   Don't know   ← page 110

I am sorry but you are wrong. It is difficult to keep a halide disc dry and a low melting alcohol would only make matters worse.

Go back to ← page 115 and choose another alternative.

You have chosen correctly. There would be a change in the spectrum. At low concentration only one band would be seen, but at higher concentration hydrogen-bonding would occur resulting in two bands.

Part 6 starts on → page 124.

Part Six

QUANTITATIVE ANALYSIS

QUANTITATIVE analysis is usually carried out on solutions which are contained in cells.

The interference fringe method is generally used to find the thickness of a cell once it has been made (Fig. 6.1).

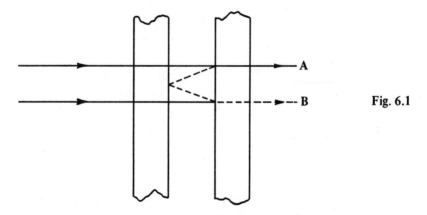

Fig. 6.1

When the beam traverses the empty cell, interference occurs between the directly transmitted beam A and the light reflected as shown by the internal surfaces of the cell to form beam B. If the light is near normal incidence and the cell thickness is t, then the path difference between the two beams is $2t$. When $2t$ is an integral number of wavelengths the beams A and B reinforce each other and the transmission is maximum, when $2t$ is a half-integral number of wavelengths the beams interfere and the transmission is a minimum.

A clean polished cell is placed in the spectrometer in the normal sample position. The pen is adjusted to be at a convenient position on the paper. The scan of the spectrometer is started and the interference fringe pattern is drawn out by the pen. Now, to calculate the thickness of the cell. The following method will only apply if the paper is calibrated linearly in wavenumbers.

Two arbitrary positions are chosen v_1 and v_2 cm^{-1} on the pattern and the number of interference fringes n between v_1 and v_2 allow the thickness to be calculated because they are related by the following equation:

$$t = \frac{n}{2} \times \frac{1}{v_1 - v_2}$$

where t is in cm

Example from Fig. 6.2 pattern (a)

$$v_1 = 1875$$
$$v_2 = 1030$$
$$n = 5$$
$$t = \frac{5}{2} \times \frac{1}{1875 - 1030}$$
$$= 0.0029 \text{ cm}$$

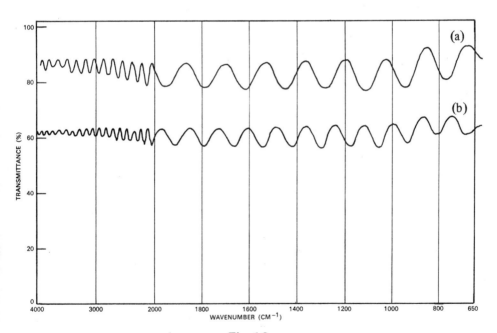

Fig. 6.2

Question
Calculate the thickness of the cell which gave pattern (b).

1. 0.0041 cm → page 139

2. 0.0082 cm → page 135

3. 0.0050 cm → page 132

The cell thickness must be expressed in centimetres for this calculation. Remember 1 μm $\equiv 10^{-4}$ cm

e.g. 100 μm $\equiv 100 \times 10^{-4}$ cm
$\equiv 0.01$ cm

Now re-calculate your cell thickness (260 μm $\equiv$? cm).

In the light of your answer, go back to → page 137 and choose another alternative.

I am sorry but you are wrong. You have either made an arithmetical mistake in your calculation or you have not used two frequencies separated by a whole number of waves. Please check your calculation.

Go back to → page 139 and choose the correct alternative.

back ref. page 130

You are wrong. You appear to have forgotten that the absorbance is a logarithmic function. Remember when you calculate the absorbance of a band the formula to use is $\log_{10} AC/BC$. When you have worked out this for the two peaks R and S then you divide these two numbers.

In the light of this information return to → page 130 and recalculate your answer.

Do you remember the formula for this?

$$\text{Absorbance} = \log_{10} I_0 / I$$

On an infrared chart the method of calculating the absorbance is as follows:
(see Fig. 6.8)

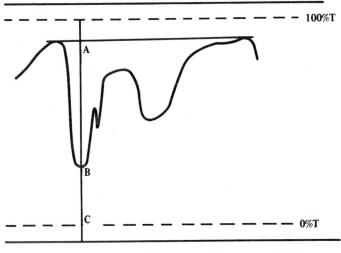

Fig. 6.8

Measure the distances AC = I_0
          BC = I

∴ Absorbance = $\log_{10}$ AC/BC

In the light of your answer, go back to → page 136 and choose another
alternative.

back ref. page 139

You are correct. The cell thickness is calculated from the interference fringe pattern as follows:

$$\lambda_1 = 13.5$$
$$\lambda_2 = 5.2$$
$$n = 11$$
$$t = \frac{11}{2} \times \frac{13.5 \times 5.2}{13.5 - 5.2}$$
$$= 47\ \mu m$$

When the cell is filled with an organic liquid, absorption bands occur and the interference pattern disappears. The intensity of an absorption band is measured using a term known as OPTICAL DENSITY (OD) but a better name is the ABSORBANCE.

OPTICAL DENSITY = ABSORBANCE

If one wishes to calculate the absorbance of an infrared absorption band from a chart which is calibrated in percentage transmittance as shown (Fig. 6.4), the correct method is as follows:

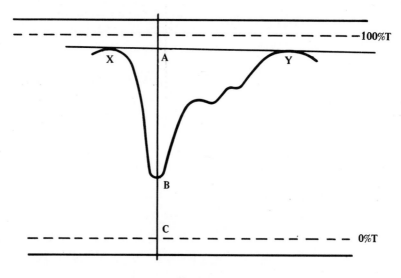

Fig. 6.4

Draw the background of the spectrum by joining the two high points X and Y. Draw a vertical line through the absorption band centre i.e. line ABC. The following conditions now apply:

I_0 is the intensity of the radiation at point A
I is the intensity of the radiation at point B

or, in other words:

I_0 = distance AC
I = distance BC

The expression for absorbance is:

$$\text{absorbance} = \log_{10} I_0/I$$
$$\text{thus absorbance} = \log_{10} AC/BC$$

and the fraction AC/BC can be ascertained by measurement of the chart. In a large number of quantitative determinations it is preferable to use the ratio of two peaks in a spectrum. This eliminates the need to measure the thickness of the sample.

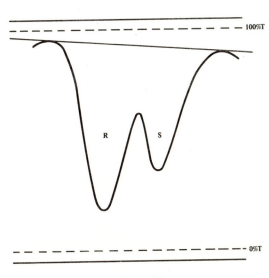

Fig. 6.5

Question
Two peaks R and S are shown in Fig. 6.5. What is the ratio of the absorbances (optical densities) of the peaks R/S?

1.   1.36   → page 133

2.   1.75   → page 136

3.   2.01   ← page 128

4.   None of these   → page 138

I am sorry but you are wrong. You have either made an arithmetical mistake in your calculation or you have not used two wavelengths separated by a whole number of waves. Please check your calculation.

Go back to ← page 125 and choose the correct alternative.

You are wrong. You have not calculated the absorbance of the absorption bands correctly. Remember that the absorbance of an absorption band is a logarithmic function and not a linear function.

Re-read ← page 130 and then attempt the question again.

Sorry your answer is incorrect, but where did you make your mistake?

(i) Did you calculate the absorbance to be
        a) 8.3? If so, turn to ← page 129.
        b) 0.83? Correct. Check parts (ii) and (iii).
        c) some other number? If so, turn to ← page 129.

(ii) Did you calculate the concentration of the solution to be
        a) 0.2 mole/l? Correct. Check part (iii).
        b) 20 mole/l? If so, turn to → page 138.
        c) some other number? If so, turn to → page 138.

(iii) Did you calculate the thickness of the cell to be
        a) 0.026 cm? If so, return to → page 137 and check your calculation.
        b) 0.260 cm? If so, turn to ← page 126.
        c) some other figure? If so, turn to ← page 126.

In the light of this information go back to → page 137 and choose the correct alternative.

I think you have forgotten to divide by 2. The formula is:

$$t = \frac{n}{2} \times \frac{1}{v_1 - v_2}$$

If you did divide by 2 you counted the waves incorrectly, ∿ is a wavelength, so pick v_1 and v_2 to be either at a crest or at a trough and count the number of crests or troughs between the two frequencies.

In the light of this information go back to ← page 125 and choose the correct alternative.

back ref. page 131

You are correct. The absorbance ratio of the two peaks is 1.75 calculated from:

Absorbance of R $\qquad \log_{10} \frac{56}{11} = 0.705$

Absorbance of S $\qquad \log_{10} \frac{54.5}{21.5} = 0.404$

$$\frac{\text{Absorbance of R}}{\text{Absorbance of S}} = \frac{0.705}{0.404} = 1.75$$

For most compounds if you double the thickness of a sample you double the absorbance of the band obtained. This is also true for solutions; if you double the concentration of a solution you also double the absorbance of the band.

A sample where this is true is said to obey Beer's law which simply states that *the absorbance of a band is directly proportional to the concentration of absorbing material for a given thickness of sample*

i.e. absorbance $\propto$ concentration $\times$ thickness

or $\qquad \log_{10} I_0/I \propto ct$

where t is the thickness of the sample in centimetres, c is the concentration in moles/litre.

$$\log_{10} I_0/I = \epsilon ct$$

where ϵ is a constant called the extinction coefficient.

Now let us see how to use this expression. By rearranging, we get

$$\epsilon = \frac{1}{ct} \log_{10} \frac{I_0}{I}$$

or $\qquad \epsilon = \frac{1}{ct} \times (\text{absorbance})$

We already know how to calculate absorbances (if you are still in doubt read page 130 and then proceed).

Let us assume that we have 5 gm of a solid whose molecular weight is 60 dissolved to give 100 ml of a solution.

We thus have 50 gm/l (Note 60 gm/l = 1 mole in this case)
Thus we have 50/60 mole/l
$\qquad$ or 0.833 mole/l.

This solution is placed into a 200 μm cell and the band obtained has an absorbance of 0.333.

$$\text{Thus} \quad \epsilon = \frac{0.333}{0.833 \times 200} \times 10^4$$

$$\epsilon = \frac{0.333}{166.6} \times 10^4$$

$$\epsilon = 20$$

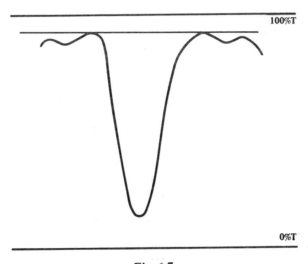

Fig. 6.7

Question
Calculate the extinction coefficient of the compound whose absorption band is shown in Fig. 6.7 given that the molecular weight is 100, the concentration is 2% w/v solution and the cell thickness is 260 μm.

1. 160   → page 143

2. 1600   → page 140

3. Some other number   ← page 134

Let us take a look at another example of the calculation of absorbance. Consider the band shown below (Fig. 6.6).

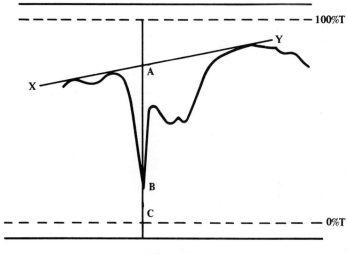

Fig. 6.6

The steps to be taken are as indicated below:

1. Draw in a line XY touching the tops of the band (this line should be near to the 100 per cent transmission line).

2. Draw a second line AC perpendicular to the 100 per cent transmission line through the bottom of the band.

3. Absorbance = $\log_{10}$ AC/BC, so measure the length of AC in centimetres and divide this by the length BC, also in centimetres. The logarithm of the result obtained is the absorbance of this band.

Calculate the absorbance of the band shown above. Your answer should be 0.67.

Now return to ← page 130 and work out the original question.

You are correct. The answer is 0.0041 cm and should have been obtained as follows:

$$t = \frac{5}{2} \times \frac{1}{1550 - 940}$$
$$= 0.0041 \text{ cm}$$

If the spectrometer is linear in wavelength (μm) a different pattern is observed which gets broader as the wavelength increases (see Fig. 6.3). The equation which allows the thickness to be calculated on this type of instrument is:

$$t = \frac{n}{2} \times \frac{\lambda_1 \times \lambda_2}{\lambda_1 - \lambda_2}$$

where λ_1 and λ_2 are the two arbitrary positions expressed in micrometres. t is expressed in μm and n is the number of waves between the two positions λ_1 and λ_2.

Example from the Fig. 6.3 pattern (a)

$$\lambda_1 = 12.0$$
$$\lambda_2 = 8.4$$
$$n = 7$$
$$t = \frac{7}{2} \times \frac{12.0 \times 8.4}{12.0 - 8.4}$$
$$= 98 \ \mu\text{m}$$

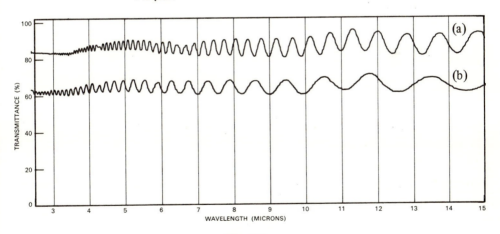

Fig. 6.3

Question
Calculate the thickness of the cell which gave pattern (b)

1.   36 μm   ← page 127

2.   47 μm   ← page 130

3.   94 μm   → page 141

Sorry your answer is incorrect, but where did you make your mistake?

(i) Did you calculate the absorbance to be
       a) 8.3? If so, turn to ← page 129.
       b) 0.83? Correct. Check parts (ii) and (iii).
       c) some other number? If so, turn to ← page 129.

(ii) Did you calculate the concentration of the solution to be
       a) 0.2 mole/1? Correct. Check part (iii).
       b) 20 mole/1? If so, turn to ← page 138.
       c) some other number? If so, turn to ← page 138.

(iii) Did you calculate the thickness of the cell to be
       a) 0.026 cm? Correct. Return to ← page 137 and check your calculatic
       b) 0.260 cm? If so, turn to ← page 126.
       c) some other figure? If so, turn to ← page 126.

In the light of this information go back to ← page 137 and choose the correct alternative.

I think you have forgotten to divide by 2. The formula is:

$$t = \frac{n}{2} \frac{\lambda_1 \times \lambda_2}{\lambda_1 - \lambda_2}$$

If you did divide by 2 you counted the waves incorrectly. $\smile\frown$ is a wavelength so pick λ_1 and λ_2 to be either at a crest or at a trough and count the number of crests or troughs between the two wavelengths.

In the light of this information go back to ← page 139 and choose the correct alternative.

You are wrong. The concentration of the sample has to be expressed in *moles per litre (mole/l)*. The method of expressing a concentration in gram molecules per litre is as follows:

e.g. concentration = 5% w/v
molecular weight = 250

∴ 5 gm of sample is dissolved in 100 ml of solvent
50 gm of sample will be dissolved in 1000 ml of solvent.

The fraction of the molecular weight dissolved in 1 litre is

$$\frac{50}{250} = \frac{1}{5} = 0.2$$

∴ the concentration of the sample in the solvent is 0.2 mole/l.

Re-read ← page 136 and then attempt the question again.

You are correct. The calculation you should use is:

$$\epsilon = \frac{1}{20/100 \times 260/10^4} \times \log \frac{58}{8.5}$$

$$= \frac{10^4 \times 100 \times 0.83}{20 \times 250}$$

$$\therefore \epsilon = 160$$

Part 7 starts on → page 144.

Part Seven

FAULT RECOGNITION

In this part the pages are arranged in numerical order, and can be read in the same way as a normal text book.

INSTRUMENTAL faults are the ones which are the most easily rectified when the sample has been prepared. The faults which arise in the sample preparation, however, take more time to rectify and therefore they should be quickly recognised before too much instrument time is wasted running bad samples. If the sample is too thick or too thin bad spectra will be obtained and probably important diagnostic detail will be missed. These types of sample should be quickly recognised on the first fast scan through the wavelength range, before the sample is run. These faults occur most frequently when you are running discs and mulls of solids.

A spectrum of a KBr disc which contains too much sample is shown in Spectrum 1. Notice how most of the absorption bands have flats at their bottom end and how wide the absorption bands are. A good spectrum usually shows the shape of all of its absorption bands but has its strongest bands coming to 99% absorption. This means that the pen should travel from 5% absorption to 99% absorption as the spectrum is scanned and not spend a large amount of time at the lower extremity. Spectrum 2 shows the same sample run as a nujol mull but this time the mull is too thin. The only clear absorption bands in this spectrum are those due to nujol and the maximum sample absorption band only comes to 70%. Spectrum 3 shows how a good spectrum should appear.

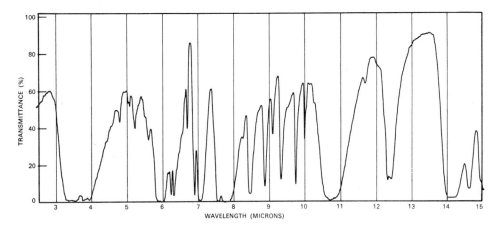

Spectrum 1

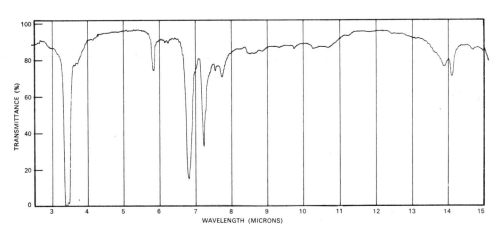

Spectrum 2

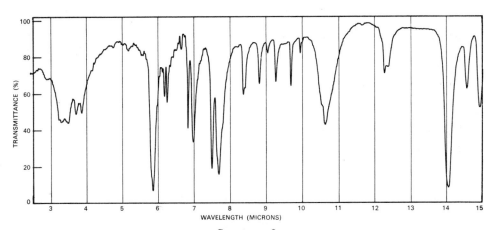

Spectrum 3

K

Solid samples which absorb water from the atmosphere or which have been washed with wet solvents and not thoroughly dried give rise to very poor infrared spectra when run as a halide disc. The reason for this is that the halide is also hygroscopic and so the water is strongly bound into the disc. The infrared spectrum of water is very strong and this dominates the spectrum of the sample. An opaque disc is usually formed when the sample is wet because water tends to stop the complete sintering of the halide when it is pressed. A large amount of scattered light gives a sloping background. The spectrum which one obtains from a wet sample is shown in Spectrum 4. Spectrum 5 shows a correctly run trace of the same sample.

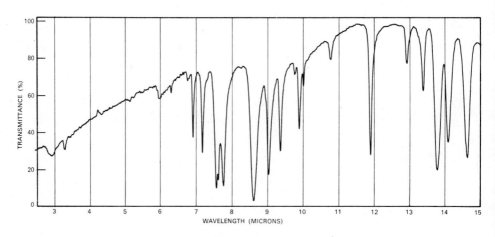

Spectrum 4

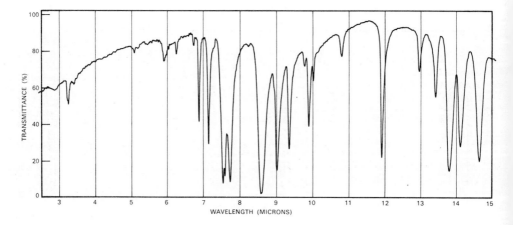

Spectrum 5

The grinding of a solid sample is important.

A badly ground sample is characterised by two features and is therefore easily recognised. The sample spectrum has asymmetric absorption bands, that is, a large number of absorption bands similar to those shown in Fig. 7.1. The spectrum is also very difficult to attenuate because it has a rising background i.e. the tops of the absorption bands lie on a curve as shown. Spectrum 6 shows a badly ground sample made up into a disc.

Fig. 7.1

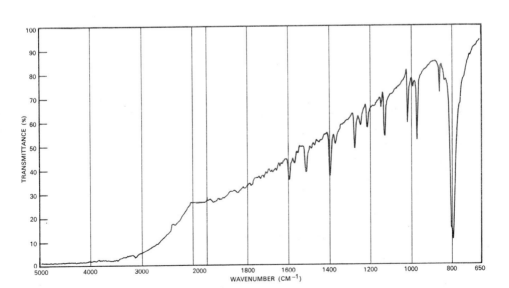

Spectrum 6

A fault which can occur when low boiling liquids (<80°C) are examined is that the liquid may evaporate as the spectrum is scanned. This arises from the fact that the source of the spectrometer emits a large amount of heat and can warm up a sample to approximately 50°C during a slow scan. This warming of the sample increases its vapour pressure and can cause bubbles to form in a badly filled or badly sealed liquid cell. Care should always be taken to see that liquid cells are completely full and sealed when low boiling liquids are being examined. Spectrum 7 shows the correctly run spectrum of a low boiling liquid, whilst Spectrum 8 shows a spectrum that has partially evaporated from its cell.

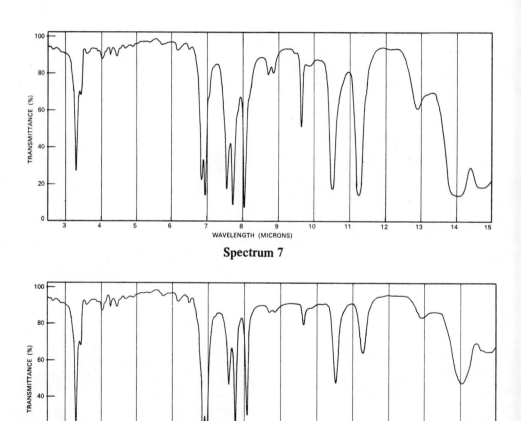

Spectrum 7

Spectrum 8

When examining liquid films or mulls one must ensure that the sample covers the whole of the entrance slits, otherwise light which has not passed through the sample will enter the spectrometer, (Fig. 7.2).

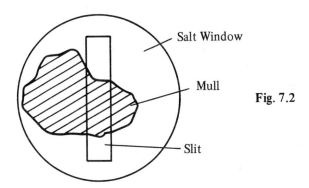

Fig. 7.2

If you present a sample such as that shown in Fig. 7.2 to the spectrometer the shape of the spectrum will be similar to that shown in Spectrum 9. Notice all the strong bands have flattish ends and do not extend to the bottom of the chart.

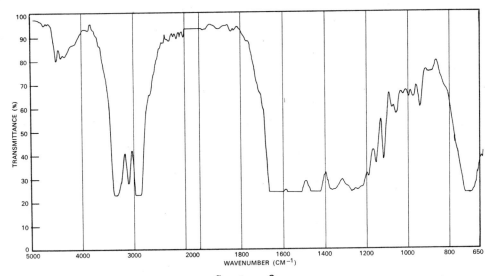

Spectrum 9

Question
Name the faults which are to be found in the following spectra.

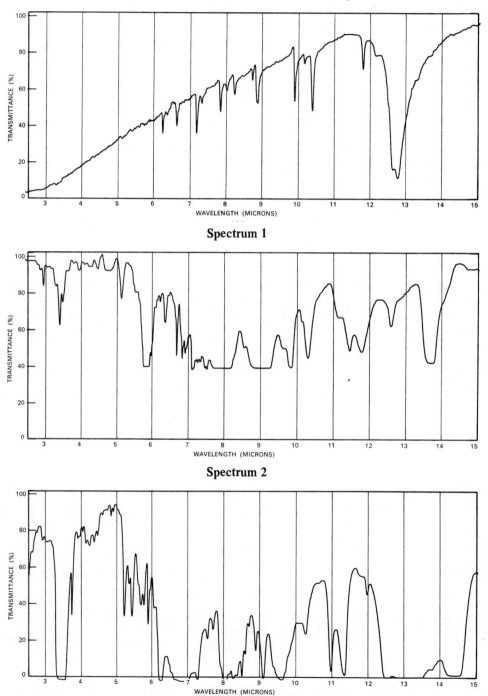

Spectrum 1

Spectrum 2

Spectrum 3

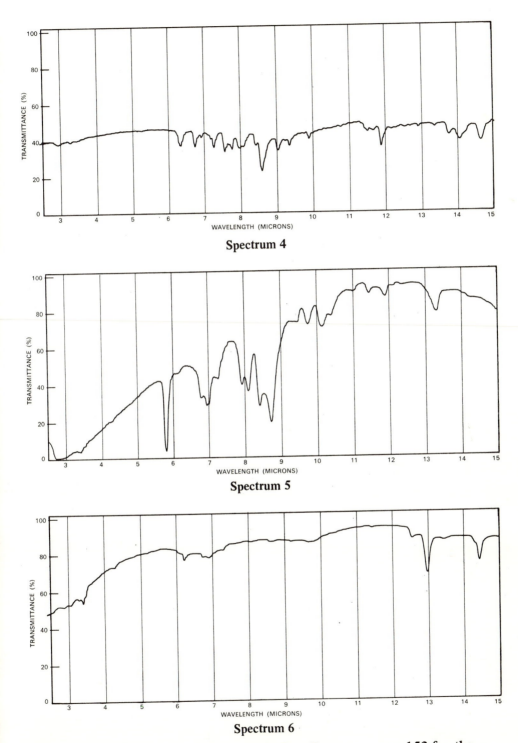

Spectrum 4

Spectrum 5

Spectrum 6

Please answer the question before continuing. Turn to → page 152 for the correct answers.

Answers
1. Badly ground sample

2. Sample not covering the slits completely

3. Too thick a sample

4. Too thin and not attenuated correctly (instrumental fault)

5. Wet sample

6. Too thin a sample

If you were wrong in some of your answers and wish to check them please return to ← page 144. If you were correct, please continue by turning to Part 8 which starts on → page 153.

Part Eight

INTERPRETATION OF SPECTRA

Now THAT the causes of absorption bands and the faults which occur have been illustrated, let us continue with the interpretation of a spectrum.

Although absorption bands are characteristic of the molecule as a whole, it is nevertheless a useful approximation to consider that molecular vibrations are localised in particular *functional groups* e.g. CH_2, CH_3, $C=O$, $C\equiv N$ etc. The absorption bands arise from the vibrations of functional groups and are almost independent of the rest of the molecule. This allows one to relate absorption band position with a particular functional group and to tabulate these relationships. The tables showing the positions where the functional group absorptions occur, are called *correlation tables*. The intensity of the absorption bands are also shown on good correlation tables.

Question
How are infrared correlation tables and functional groups related?

1. A correlation table is a list of functional groups. → **page 155**

2. A correlation table relates the position of an infrared absorption band with the functional group causing the absorption. → **page 159**

3. A correlation table relates the intensity of an absorption with the functional group. → **page 167**

A C=N stretch is not shown as occurring above 1700 cm⁻¹ on the correlation chart and as the absorption band in question is above 1700 cm⁻¹ you have assigned it incorrectly.

Re-read → page 159 and then select the correct alternative.

Although a list of functional groups occur in correlation tables your answer is wrong. Anyone can write out a list of functional groups but what are you relating this list to?

Re-read ← page 153 and select the correct alternative.

Well done. You found that the C=O stretch was the only one of the alternatives which occurred above 1700 cm⁻¹ in the correlation table. The spectrum was that of acetone.

The method for starting to recognise characteristic group frequencies is described on the next few pages. Always start at the left-hand side and look across the spectrum as follows.

Spectrum 1
This is a spectrum of cyclohexane and contains only CH_2 vibrations. The pair of bands at 2920 cm⁻¹ and 2850 cm⁻¹ are due to CH_2 stretching vibrations and are written $\nu(CH_2)$. As both symmetric and asymmetric vibrations are infrared-active the band occurs as a doublet. The absorption band at 1460 cm⁻¹ is a CH_2 deformation band and is written $\delta(CH_2)$. It is the CH_2 scissors.

Spectrum 2
This is a spectrum of n-decane and contains both CH_2 and CH_3 groups. As you can see by comparing this with Spectrum 1 all the CH_2 bands are unchanged but there is an additional band at 1385 cm⁻¹. This is characteristic of CH_3 groups and is the symmetric deformation band $\delta_s(CH_3)$. Note also that the $\nu(CH_3)$ shows as a doublet in the $\nu(CH_2)$ region.

Spectrum 3
This spectrum is given by the compound 2,4-dimethyl pentane.

$$\begin{array}{ccc} CH_3 & & CH_3 \\ | & & | \\ CH-CH_2-CH \\ / & & \backslash \\ CH_3 & & CH_3 \end{array}$$

As you can see it has in its structure two methyl groups attached to one carbon atom. This is a gem-dimethyl compound and gives rise to the symmetric doublet shown in the spectrum.

Spectrum 4
This is a spectrum of a branched hydrocarbon 2,2-dimethyl hexane and it contains a tertiary butyl group

$$\begin{array}{c} CH_3 \\ | \\ CH_3-CH_2-CH_2-CH_2-C-CH_3 \\ / \\ CH_3 \end{array}$$

This causes the methyl deformation band to become more intense and complex. The doublet formation of this band usually indicates two or more methyl groups attached to one carbon atom and for three methyls it is asymmetric. The single methyl absorption shows in between the absorptions due to the tertiary butyl group. **Now compare and study all four spectra and then try to answer the question on → page 158.**

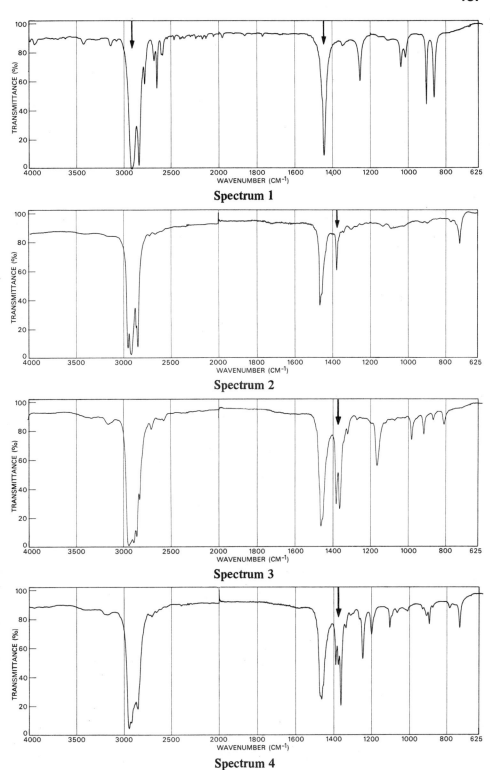

Spectrum 1

Spectrum 2

Spectrum 3

Spectrum 4

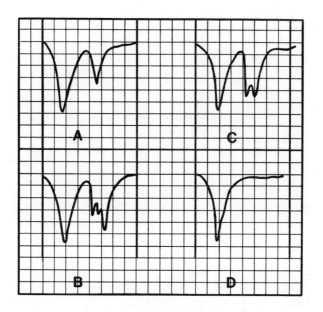

Fig. 8.3

Question

The spectra on the previous pages show the absorption bands which one may expect to see in alkanes. The region which indicates the branching is 1500 to 1300 cm^{-1}. Fit the three structures given below to the small portions of the spectra shown (Fig. 8.3).

1.
$$\begin{array}{c} CH_3 \\ | \\ H_3C-H_2C-C-CH_2-CH_3 \\ | \\ H \end{array}$$
3-methyl pentane

2.
$$\begin{array}{c} H_3C-HC-HC-CH_3 \\ | \quad | \\ CH_3 \; CH_3 \end{array}$$
2,3-dimethylbutane

3.
$$\begin{array}{c} CH_3 \\ | \\ CH_3-C-CH_2-CH_2-CH_3 \\ | \\ CH_3 \end{array}$$
2,2-dimethylpentane

Please turn to → page 163 for the answers.

You are correct. Correlation tables relate the position of infrared absorption bands with the functional group causing the absorption.

Correlation tables should be used intelligently. Assignments should not normally be made on a single absorption band unless verification from another part of the spectrum can be made. However, if an absorption band is found which lies in the 2700 to 2000 cm⁻¹ region of the spectrum and the method of preparation or the elements present in the compound are known, it is possible to assign this absorption band without verification from any other region of the spectrum. This is because there are only a few select fundamental vibrations which occur in this region. *Absorption band shapes and intensities are also important and assignments should not be made on the positions alone.*

A way of learning how to use correlation tables is to obtain a series of spectra of similar compounds which contain one functional group other than C–H. For example, look at a number of compounds which contain O–H, C=O or C–O–C=O. Look at one series and examine the strong absorption bands, try to remember their shapes and the positions in which they occur. Now look at the correlation tables and knowing what the compounds are, assign the strong absorption bands. It is best to start this way because you will notice that in the correlation tables a large number of different vibrations overlap (Fig. 8.1 pages 160 and 161). It is usually the shape and intensity which enables you to distinguish between these overlapping absorption bands.

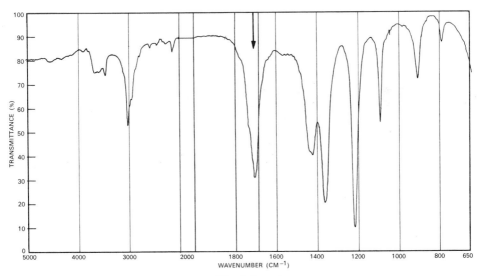

Fig. 8.2

Question

In the spectrum shown (Fig. 8.2) can you identify the absorption band marked?

1. C=O stretch ← page 156

2. C=N stretch ← page 154

3. C=C stretch → page 173

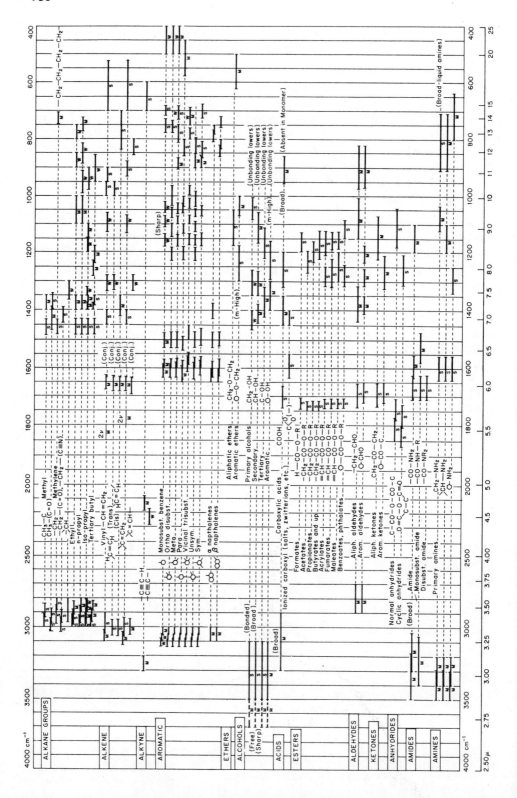

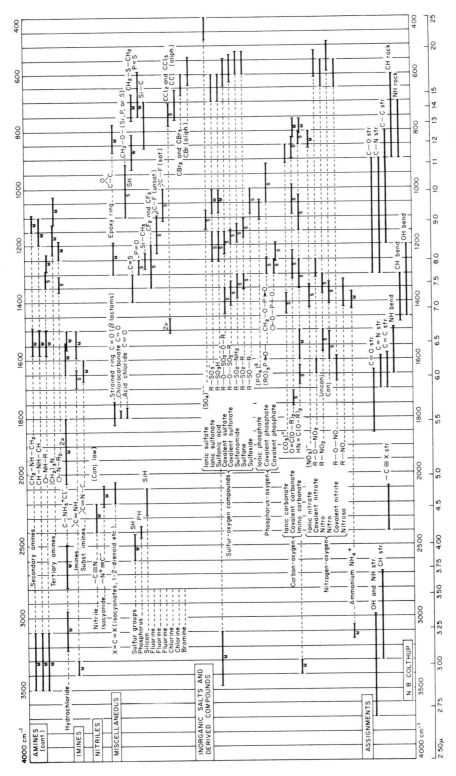

Fig. 8.1 Correlation Tables. *(Courtesy of N. B. Colthup).* Originally published in *Applications of Absorption Spectroscopy of Organic Compounds* by J. R. Dyer, Prentice-Hall, New Jersey, 1965.

M

back ref. page 172

The $\nu(C-O)$ is at 1210 cm^{-1}. This is a higher frequency than those of the previous spectra of alcohols and it has been pointed out that the fewer the hydrogens on the carbon attached to the O–H group, the higher the frequency. If you remove hydrogens from a secondary alcohol and replace them with carbon, what type of alcohol would you obtain?

Think about the question, when you can answer it you should be able to answer correctly the question on → page 172.

Answers

1.   a

2.   c

3.   b

Please note: When two CH_3 groups are bonded to one carbon **two** symmetrical CH_3 deformations will occur. As these can vibrate either in-phase or out-of-phase the $C-CH_3$ bond will split. Since the coupling is weak this splitting will not be large. Spectrum D is that of a cyclic compound containing no CH_3 groups.

Please continue by going to → page 170.

Correct. The spectrum was that of a tertiary alcohol. The ν(C–O) has moved to a frequency higher than that found in isopropanol and the δ_s(CH$_3$) is also very strong, showing that the compound contains a larger number of CH$_3$ groups. This helps to confirm that the spectrum is that of tertiary butanol.

Spectrum 9
This is another type of compound. You start at the left hand side again and the strong bands below 3000 cm^{-1} indicate aliphatic CH groups only. The next large band which occurs is marked $\downarrow$. Looking at the correlation tables you will see that this occurs in a region marked for ν(C=O). The compound is carbonyl containing and is a ketone. It is, in fact, cyclohexanone. Notice the shape and intensity of the bands. The other bands are fingerprint absorptions of the compound.

Spectrum 10
This is the spectrum of methyl ethyl ketone. Start at the left hand side and you see the band is below 3000 cm^{-1}, indicating an aliphatic hydrocarbon. The next strong band is at 1720 cm^{-1} and is characteristic of a ketone. Now look at the deformation bands; notice how strong the band is which occurs at 1375 cm^{-1}. This band, if you remember, was δ_s(CH$_3$) in the hydrocarbon spectra. It is the same absorption now but it is intensified because it is adjacent to the C=O and also because it contains a relatively large number of CH$_3$ groups.

Spectrum 11
The final spectrum in this series is that of methyl iso-butyl ketone. Notice the positions of the bands marked in the spectra and the fact that they have similar shapes and intensities. The δ_s(CH$_3$) has increased in intensity relative to the other CH bands. This is because the compound contains extra CH$_3$ groups and is in agreement with the previous alkyl molecules.

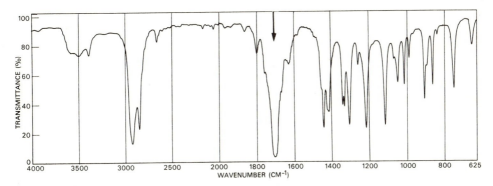

Spectrum 9

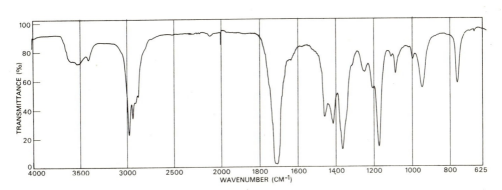

Spectrum 10

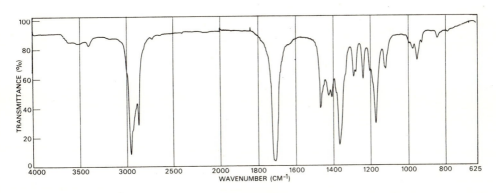

Spectrum 11

Now compare and study all three spectra and then try to answer the question
on → page 166.

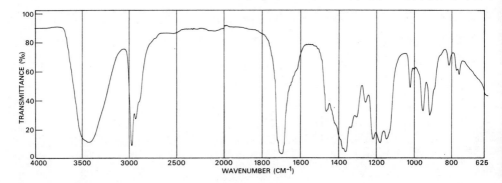

Spectrum 12

Question

Can you identify the functional groups other than $(CH)_n$ which are present in the single compound shown in Spectrum 12? How many functional groups are there?

1. one → page 178

2. two → page 174

3. three → page 169

Your answer is incorrect. The intensity of an absorption band and its relationship with the functional group is normally only a part of a correlation table.

Re-read ← page 153 and select the correct alternative.

The ν(C–O) is at 1210 cm^{-1}. This is a higher frequency than those of the previous spectra of alcohols and it has been pointed out that the fewer the hydrogens on the carbon attached to the O–H group, the higher the frequency. If you remove hydrogens from a secondary alcohol and replace them with carbon, what type of alcohol would you obtain?

Think about the question, when you can answer it you should be able to answer correctly the question on → page 172.

I am sorry but you are wrong. You have found the three absorption bands at 3400 cm^{-1}, 1706 cm^{-1} and 1180 cm^{-1} but two of these belong to the same group, in that they both have to be present in a compound which contains this group.

Go back to ← page 166 and choose another alternative.

Spectrum 5

When you look at this spectrum you will see that a compound has been chosen containing two strong bands in addition to the spectra of the hydrocarbons whose bands were marked ↓ . From the correlation tables you will find that the first band at 3300 cm^{-1} may be either N$^+$-H, N—H bonded, N—H, NH$_2$ free, OH bonded or C≡C—H. Let us try to eliminate the alternatives.

Nitrogen containing compounds also have bands in other positions which are given in the correlation tables. N—H bend in-plane occurs between 1650 and 1500 cm^{-1} and an absorption does occur here. An out-of-plane bend occurs between 900 and 700 cm^{-1} but we have no definite absorption in this region. In a nitrogen containing compound there must also be a ν(C—N); this is shown to occur between 1300 and 900 cm^{-1}. A strong band is marked at 1030 cm^{-1} Of the other two alternatives C≡C—H must have a ν(C≡C) and this would be found at ~2200 cm^{-1}. No band is observed in this region so there is no C≡C—H in the compound. The ν(OH) bonded would occur in an alcohol and such a compound would also contain a ν(C—O). This occurs between 1300 and 900 cm^{-1} and a band is found at 1030 cm^{-1}. The evidence therefore suggests that the compound is an alcohol or an amine but the evidence for an amine is incomplete because there is no N—H out-of-plane band. However, all of the bands may be assigned to an alcohol. The spectrum is, in fact, that of methanol, but to distinguish absolutely between an amine and an alcohol other properties should also be considered such as smell molecular weight and boiling point. By considering these other properties at the same time as finding the alternatives from the spectrum, your answer will be more reliable. The series of spectra with the absorption bands marked ↓ are those of aliphatic alcohols.

Spectrum 6

This is a spectrum of n-propanol. Starting from the left hand side of the spectrum and looking at strong bands we see a band at 3300 cm^{-1}; this is the ν(OH). We confirm that it is an alcohol by looking for the ν(C—O). This is given by the strong band at 1070 cm^{-1}. Now look back to the left hand side of the spectrum and notice the strong band below 3000 cm^{-1}. This is aliphatic ν(CH)$_n$, both asymmetric and symmetric bands showing.

Spectrum 7

This is a spectrum of a secondary alsohol, isopropanol. Again we see the same pattern of strong bands. The strong band at 3300 cm^{-1} is the ν(OH) and the strong bands centred at 1140 cm^{-1} are the ν(C—O). Have you noticed that as you go from methanol H_3C—O—H → R—H_2C—OH → R_2HCOH the position of the strong band is moving to a higher frequency? This is another useful piece of information which may be obtained from the infrared spectrum. Besides saying that the compound is an alcohol one can generally say what type of alcohol it is i.e. primary, secondary or tertiary. Notice also in the spectrum the ν(CH) bands and the δ(CH); notice how the δ_s(CH$_3$) has increased relative to the δ_s(CH$_3$) in spectrum 6. This indicates that the compound contains more than one CH$_3$ and helps to confirm that the sample is a secondary alcohol.

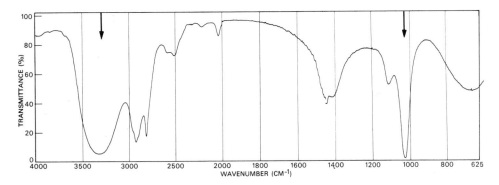

Spectrum 5

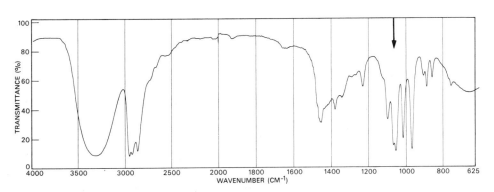

Spectrum 6

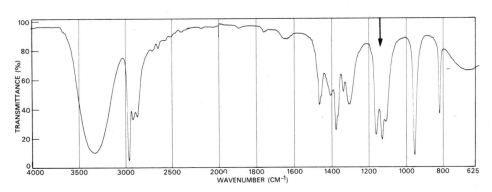

Spectrum 7

Now compare and study all three spectra and then try to answer the question on → page 172.

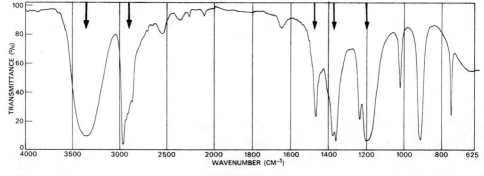

Spectrum 8

Question
Spectrum 8 is that of an aliphatic alcohol. What type of alcohol is this? The bands which you have to consider have been marked.

1. Primary ← **page 162**

2. Secondary ← **page 168**

3. Tertiary ← **page 164**

A C=C aliphatic absorbs very close to 1700 cm^{-1} but it does not go above this wavenumber. You did not read the frequency of the absorption band carefully enough. Re-check your assignment.

Go back to ← page 159 and choose another alternative.

You are correct, there are two functional groups C–OH and C=O present in the compound (diacetone alcohol), the structure of which is

$$CH_3-\overset{\overset{\displaystyle O}{\|}}{C}-CH_2-\overset{\overset{\displaystyle CH_3}{|}}{\underset{\underset{\displaystyle CH_3}{|}}{C}}-OH$$

Spectra 13, 14, 15

This is a series of spectra of other types of carbonyl compounds. Spectrum 13 has a strong absorption band at 1740 cm^{-1} and there is no strong band for ν(OH) as in the diacetone alcohol spectrum. There is, however, a strong band at 1245 cm^{-1} which is a ν(C–O). A combination of a rise in frequency of a ν(C=O) and the appearance of a ν(C–O) is indicative of esters. This spectrum is that of the ester n-propyl acetate.

Spectrum 14 is that of another aliphatic ester (n-butyl butyrate) but not an acetate. Look at the ν(C–O) which came in acetates at 1245 cm^{-1}. Now it is lower in frequency and occurs at 1180 cm^{-1}.

Spectrum 15 is that of acetic acid. The ν(C=O) is now lower in frequency and there is still the strong band in the ν(C–O) region. Notice now that a very broad band centred near to 3000 cm^{-1} is overlapping with the narrower ν(CH$_3$). Another useful band is the one about 930 cm^{-1}. Look at the shape and try to remember this when looking at this type of compound as it is very characteristic of COOH-containing compounds.

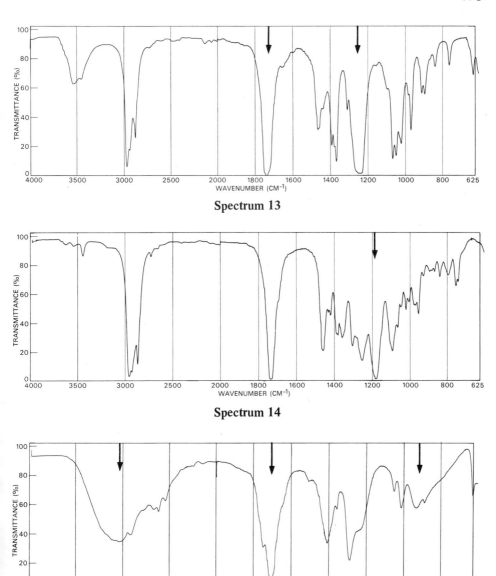

Spectrum 13

Spectrum 14

Spectrum 15

Now compare and study all three spectra and then try to answer the question
on → page 176.

176

Question

Interpret the six spectra shown and match them with the formulae given below.

1. $CH_3 - CH_2 - \overset{\overset{\displaystyle O}{\|}}{C} - OH$

2. $CH_3 - \overset{\overset{\displaystyle O}{\|}}{C} - O - CH_2 - CH_3$

3. $CH_3 - \overset{\overset{\displaystyle O}{\|}}{C} - CH_3$

4. $CH_3 - \overset{\overset{\displaystyle CH_3}{|}}{CH} - CH_2 - \overset{\overset{\displaystyle CH_3}{|}}{CH_2} - CH_3$
   $\qquad\qquad\qquad\qquad\quad \underset{\underset{\displaystyle CH_3}{|}}{}$

5. $CH_3 - \overset{\overset{\displaystyle CH_3}{|}}{\underset{\underset{\displaystyle CH_3}{|}}{C}} - OH$

6. $CH_3 - CH_2 - CH_2 - CH_2 - \overset{\overset{\displaystyle CH_3}{|}}{\underset{\underset{\displaystyle CH_3}{|}}{C}} - CH_3$

Please write down your answers and turn to → page 179.

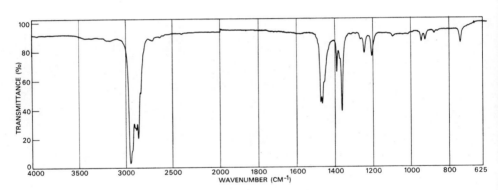

Spectrum A

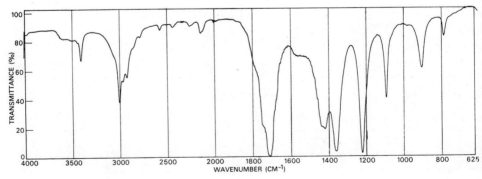

Spectrum B

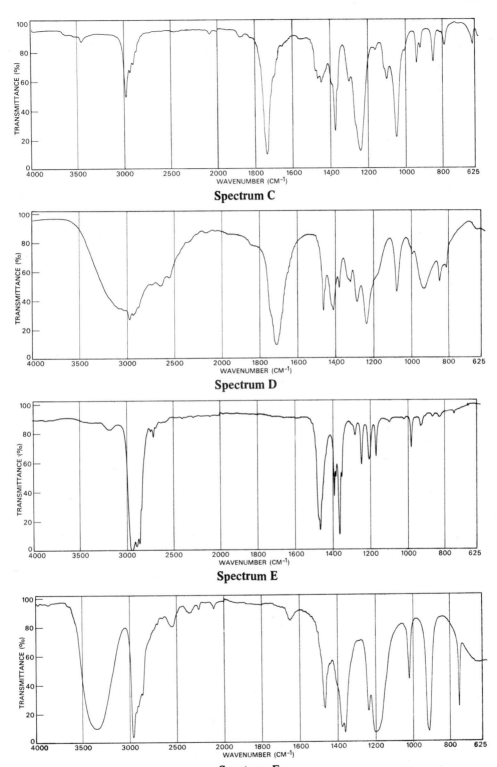

Spectrum C

Spectrum D

Spectrum E

Spectrum F

N

I am sorry but you are wrong. There are absorption bands at 3400 cm^{-1}, 1706 cm^{-1} and 1180 cm^{-1} which are characteristic of groups other than $(CH)_n$.

Go back to ← page 166 and choose another alternative.

Answers
The order of the compounds were:

Spectrum A — 6
         B — 3
         C — 2
         D — 1
         E — 4
         F — 5

If you got any wrong, look back at the section dealing with the compound and check the assignments.

If you continue to work systematically through your own spectra and look for common bands in a series and then refer to the correlation tables you will quickly progress in the recognition of infrared spectra.

180

Appendix   A

Attenuated Total Reflection

THERE WAS in the table of window materials a statement that KRS 5 was an ATR crystal material. The letters ATR stand for *attenuated total reflection*. This is the name given to a phenomenon associated with total internal reflection at a crystal surface and may be explained as follows.

When a beam of radiation enters a prism it will be totally internally reflected when the angle of incidence is greater than the critical angle, the critical angle being a function of the refractive index of the crystal material, and it is preferable if this is a fairly high number between 2 and 4. When the light is reflected all of the energy is reflected. However, the beam of radiation appears to penetrate slightly beyond the reflecting surface and then return to the prism (see Fig. 1).

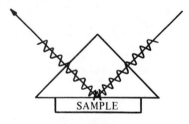

Fig. 1

When a material which selectively absorbs some of the radiation is placed in contact with the reflecting surface the beam of radiation will lose some of its energy i.e. the energy will have been attenuated.

Let us assume that the crystal is made of KRS 5 and pressed into close contact with the surface to be used to reflect the radiation, is an organic sample. This system is now placed in a suitable holder and mirror system in the sample beam of an infrared spectrometer. The sample absorbs radiations at its characteristic positions and what is drawn out on the chart is an absorption spectrum of the sample. The intensity of the radiation absorbed is dependent not on the sample thickness but on the penetration distance, which is again a function of the relative refractive indices of the crystal and sample. A single reflection may not be sufficient to give a good spectrum, the quality of which may be improved by the use of a multi-reflection crystal rather than by making the sample thicker.

Appendix   B

Sample Preparation

Form of material	Method of sample treatment
1. Gas sample	Place into an evacuated gas cell. If very low concentration, put it into a multi-reflection cell.
2. Liquid sample	Examine neat by placing in a liquid cell with a thickness of $\simeq 20\mu$m. If it is a high boiling liquid it may be placed between two windows without a spacer and the capillary film run.
3. Waxy or low melting solid	Place a window onto a cool hot plate, put onto the window a small amount of the sample, place another window on top of the sample and heat until the sample melts. Carefully squeeze the windows together but do not exert any pressure if it has not completely melted otherwise you might break the windows. Another point to watch is that you do *not* subject the windows to any thermal shock i.e. take off the hot plate and then place onto a cold bench otherwise you may break the windows.
4. Crystalline solids	Grind well and then make a mull using nujol as the dispersing agent or if you are specifically looking for aliphatic hydrocarbon bands use Kel F as the dispersing agent or grind well and then make a halide disc.
5. Polymer samples (a) films	Examine without further treatment. If too thick, examine by the ATR technique, or reduce the thickness by one of the following techniques.
(b) general	Dissolve and then cast a film either onto glass (to be removed before examination) or directly onto a halide window. Hot press into a thin film. Microtome a thin section. Cold grind dry and then treat as a crystalline sample, by preparing either disc or mull.

Appendix   C

Fault Finding

Symptoms	Fault	Remedy
Sample background rises up from near zero transmission at 2.5 μm.	Sample is wet.	Dry the KBr or nujol and check dryness of sample. If the sample is thermally stable the KBr disc may be placed in an oven at 120°C and the disc then re-pressed.
Asymmetrically shaped bands on a sloping background.	Badly ground sample.	Grind the sample and the KBr separately then grind them together. If this is unsuccessful try a mull or, if the sample has a low melting point, a melt.
Bands become weaker as the spectrum is scanned.	The sample is evaporating and hence the cell is emptying.	Stopper cell properly. If leaking continues use a new cell. If this does not solve the problem try using a gas cell and running a vapour spectrum.
Many or all bands have flat ends.	a) Sample too thick. b) If the flat ends are above 0 per cent transmission, the sample is not covering the slits.	a) Reduce sample concentration or thickness. b) Move the sample to cover the slits, or if a KBr disc, re-grind the sample to reduce its particle size.
No strong bands seen in the spectrum.	Sample is too thin or the concentration is too low.	Increase sample thickness or concentration.

Trace is very unsteady (noisy).	The gain is too high.	Reduce gain until set correctly. If trace is still noisy widen slits then re-adjust the gain.
Pen drawing steps.	The gain is too low.	Increase the gain. If this leads to noise, widen the slits.
No structure to the bands.	Poor resolution.	Reduce the slit width and increase the gain to compensate for this.

A Guide to Further Reading

H. A. Szymanski, *Interpreted Infrared Spectra,* Vols. 1, 2 and 3, Plenum Press, New York, 1964.

K. Nakanishi, *Infrared Absorption Spectroscopy — Practical,* Holden-Day, San Francisco and Nankodo Co., Tokyo, 1962.

R. G. J. Miller and B. C. Stace (Eds.), *Laboratory Methods in Infrared Spectroscopy (2nd Edition),* Heyden & Son, London, 1972.

A. D. Cross, *Introduction to Practical Infrared Spectroscopy (3rd Edition),* Butterworths, London, 1969.

L. J. Bellamy, *The Infrared Spectra of Complex Organic Molecules (2nd Edition),* Methuen, London and Wiley, New York, 1958.

J. H. van der Maas, *Basic Infrared Spectroscopy (2nd Edition),* Heyden & Son, London, 1972.

J. R. Dyer, *Applications of Absorption Spectroscopy of Organic Compounds,* Prentice-Hall, New Jersey, 1965.

N. L. Alpert, W. E. Keiser and H. A. Szymanski, *IR Theory and Practice of Infrared Spectroscopy (2nd Edition),* Heyden & Son, London/Plenum Press, New York, 1970.

Summary and Notes

Infrared spectroscopy is probably the most extensively used spectroscopic technique because it is used for all types of samples — gases, liquids and solids. It is not limited to any one type of sample, as any compound which has covalent bonds in its composition will have a characterisable infrared spectrum. Even the window materials NaCl, KBr etc. which are ionic in character may be recognised by the point at which the limit of infrared transmittance occurs.

The infrared examination of gaseous samples is not limited to those which are normally gases at room temperature although a multi-reflection gas cell is now being used to detect and identify atmospheric pollutants such as ammonia and sulphur dioxide. The examination in a heated cell of gaseous samples, which at room temperature are normally liquids or solids, is widely practised, especially when the infrared spectrum of a sample eluting from a gas chromatograph is required to enable the chemist to identify the fraction.

Liquid samples can either be pure, or solutions of solid samples. The running of solutions is not widely used for qualitative identification purposes as even the best solvents for use in the infrared region i.e. carbon disulphide and carbon tetrachloride cannot normally be used if they are more than a fraction of a millimetre thick. Even so, many of the best solvents (alcohols, ethers, ketones) absorb much too strongly to be useful except over very limited ranges. Solid samples can rarely be examined 'in bulk' but can be prepared as pressed discs in window materials or as 'mulls' in mineral oils. Some specialised methods are, however, available to deal with the variety of solids that are presented for examination.

Infrared spectroscopy of all these types of samples is important because of the unique insight it gives into their chemical structure.

Molecules are made up of atoms linked by chemical bonds. The movement of the atoms and the bonds may be illustrated mechanically by balls and springs in motion. The movements may be divided into two components, namely the stretching and the bending of the bonds joining the atoms. The frequencies of vibration of the atoms are dependent on the masses involved e.g. C–H, C–O and the nature of the bonds e.g. C≡N, C=N in a way related to the

structure of the entire molecule and its environment. This is why each compound has a unique infrared spectrum and taking the infrared spectrum of a
compound is like taking a person's fingerprints.

Certain atomic movements give rise to bands which occur in approximately
the same position in a large variety of compounds and are only slightly affected by the rest of the molecule. These vibrations may therefore be assigned to
certain groups of atoms which are termed *functional groups*. These functional
groups of atoms have distinguishing characteristics, e.g. multiple bonds, C=O,
C≡N, C=C etc. or single bonds between dissimilar atoms e.g. C–H, C–O,
O–H etc. where the vibrations of the group do not affect the other groups in
the molecule. If the molecule contains two functional groups which are the
same and which are adjacent or nearly adjacent, certain vibrations can and do
couple together so moving the absorption band of the single group to either
a new, characteristic, position or the coupling can be such that the vibration
is no longer easily recognised (see Fig. 1).

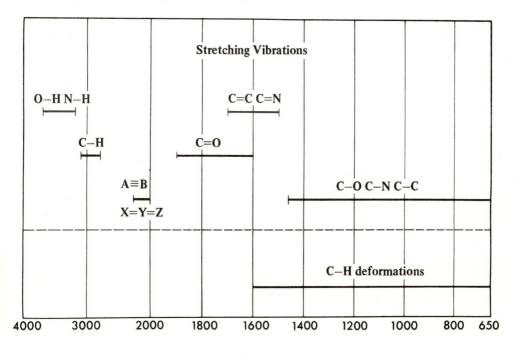

Fig. 1.

For a molecule composed of n atoms there are $3n$ degrees of freedom associated with the momentum coordinates. Of these, three will be translational
modes, and for a non-linear molecule three will be rotational modes (only
two rotational modes in a linear molecule). This means that there will be
$3n - 6$ or $3n - 5$ degrees of freedom of the molecule, each of which gives
rise to a vibration of the molecule. Certain vibrations may occur for a molecule which do not necessarily result in an infrared absorption being observed.

Some sets of vibrations are degenerate i.e. they are identical but in perpendicular directions and these multiple vibrations only result in one infrared absorption band being seen in the spectrum. The prime necessity for an infrared absorption to occur is that there must be a change in *dipole moment* in the molecule when the vibration occurs. The greater the change in dipole moment the stronger is the infrared absorption. This explains why the absorption bands of hydrocarbons are much weaker than the absorptions of groups whose components differ considerably in their electronegativities e.g. $C=O$, $C-O$, $O-H$. The different intensities of the absorptions enable the $\nu C-O$ and the $\nu C-N$ to be detected easily in a spectrum even though they occur in a region which is overlapped by the $\nu C-C$ and $\delta C-H$ absorptions.

There are no rigid rules for interpreting an infrared spectrum but certain requirements are necessary for a satisfactory result to be obtained:

1. The spectrometer should be accurately calibrated or calibration bands of polystyrene should be drawn on the spectrum to be interpreted.
2. The spectrum should be a good one.
3. The sample should be reasonably pure or its chemical history known so that the 'impurities' may be recognised.
4. Any solvents or mulling agents should be written on the chart to avoid any confusion.

A precise assignment of all the absorptions occurring in the spectrum is not feasible and should not be attempted when you are interested only in a qualitative identification of the compound. Instead, information as to the presence or absence of specific functional groups should be sought.

The absorption bands of the functional groups have been tabulated in correlation tables and this might lead one to think that these are all that are required to enable one to identify an unknown compound. This is not the whole truth because useful though they are, the unknown spectrum has still to be matched with that of a known standard for complete identification. It is because of this that there can be no substitute for experience and it is essential for beginners to examine the spectra of as many different types of compound as possible. Initially this should be done to try to fix in the mind the general appearance of a spectrum and where one would normally expect to see the absorption bands. Then one should look critically at the shapes and relative intensities of the absorption bands of series of compounds and try to relate these shapes and intensities to the assignments in the correlation tables. It helps to start looking at the spectrum as one would read a line of print in a book, starting at the first intense band or 3000 cm^{-1} for the first clue of functional groups or type of compound present. These are X$-$H stretching vibrations. Confirmation of this first tentative assignment is obtained by looking for other good characteristic absorptions of this first group in the remainder of the spectrum. One should then pass in turn to the other regions for stretching vibrations, again first looking for the strong absorptions, and confirming the groups if possible. By using a 'look and confirm' technique a list of assigned functional groups is quickly obtained plus other unidentified absorptions. This is where the experience that has been built up will help to decide whether to carry on looking at the weaker absorptions, or to start looking for confirmation, or for compounds of a similar structure in a standard commercial collection. There are over one hundred thousand standard spectra available in the literature and commercial collections. The largest single collection is The Sadtler Standard Spectra. Because of the large numbers involved it has for a long time been necessary to have mechanical aids for searching these collections. Sadtler produces a coded system called the Specfinder for the spectra in this collection. The full potential of searching the standard spectra collections is realised by using a computer. The American Society for Testing and Materials (ASTM) have produced the coded data and the computer programs which enable this to be carried out.

Infrared spectroscopy is a relatively old technique but it is also a changing one with many new developments occurring in instrumentation and data handling so that it offers much promise for the future.

188

Answers

Check your answers to the Criterion Test against those given below:

		Individual Marks	Total Marks
1.	$2.5\ \mu m$ to $15\ \mu m$ *or* $4000\ cm^{-1}$ to $650\ cm^{-1}$.		1
2.	To remove unwanted orders produced by a grating spectrometer.		4
3.	(1) Source (2) Chopper (3) Slit (4) Prism (5) Littrow mirror (6) Detector (7) Amplifier (8) Servo motor (9) Servo comb (10) Recorder.	½ *mark each*	5
4.	Any three of these window materials are accepted as correct. (a) NaCl (b) AgBr (c) KBr (d) AgCl (e) CsI (f) KCl (g) KRS.5.	*Any three 1 mark each*	3
5.	(a) Four (4) (b) Three (3) two identical (c) the vibration has to change the dipole moment of the vibrating group of atoms or molecule.	*(a) 3 (b) 3 (c) 5*	11
6a.	Any two of these mulling agents is accepted as correct. (i) Nujol (ii) Kel F (iii) hexachloro-butadiene	*Any two 1 mark each*	2
6b.	It is necessary to use mulling agents such as nujol and Kel F because there is a possibility that some of the absorption bands of the sample may be masked by those of the mulling agent if only one is used.		2
7.	The formula used in the calculation is :— $$t = \frac{n}{2} \times \frac{\lambda_1 \lambda_2}{\lambda_1 - \lambda_2}\ \mu m$$ answer (a) $98\ \mu m$ (b) $47\ \mu m$	*Correct formula 5 marks* *Correct answers 4 marks each*	13

8. $A = \log \frac{I_0}{I} = \epsilon \, cl$ *Correct formula*
5 marks

From the spectrum shown
$I_0 = 58$
$I = 8.5$ $\log \frac{I_0}{I} = 0.834$ *Correct absorbance*
3 marks

The concentration is 1 mol/l
The path length is 47×10^{-4} cm
The calculation you should use is therefore
$0.834 = \epsilon l \times 47 \times 10^{-4}$ which reduces to

$$\epsilon = \frac{0.834 \times 10^4}{1 \times 47}$$

$\epsilon \simeq 178$ *Correct answer*
4 marks   12

9. A multi-reflection gas cell would be used when
there is only a small sample available or when   *2 marks*
the concentration of the sample is very small
in an inert gas.

     2 marks   4

10. The type of table you would use would be a
Correlation table.   2

11. (a) Hot pressing a film
(b) Casting a film from a solvent   *2 marks*
each
(c) Microtoming a thin section   6

12. (a) The absorption band caused by the
stretching or bending of a functional group of
atoms vibrating with simple harmonic motion   *5 marks*
(b) The absorption occurring at $n\nu$, where ν is
the wavenumber of the fundamental absorption
band and $n = 2,3$ etc.   *5 marks*

(c) The absorption band occurring at $v_1 + v_2$ or $v_1 - v_2$ where v_1 and v_2 are the wavenumbers of the fundamental absorption bands of a function group of atoms.   *5 marks*   15

13a.   The relationship is :

$$\text{wavenumber} = \frac{1}{\text{wavelength in } \mu\text{m}} \times 10^4$$

     4

13b. $6 \, \mu\text{m}$ 2

14.   (a) C=N 1700 cm^{-1} to 1500 cm^{-1} *2 marks*
       (b) C=O 1800 cm^{-1} to 1600 cm^{-1} *2 marks*
       (c) C−OH 3500 cm^{-1} to 3300 cm^{-1} and
       1300 cm^{-1} to 1000 cm^{-1} *2 marks*   6

15a.   (i) Thermocouple (ii) Bolometer (iii) Golay or   *Any two*
       pneumatic detector.   *2 marks each*   4

15b.   (a) The change in e.m.f. or the voltage produced at the junction of two dissimilar metals or semiconductors.
       (b) The change in resistance of a heated wire   *Any two*
       causes a change in current flowing through the   *2 marks each* wire.
       (c) The expansion of a gas causes a light beam to vary in intensity in a photoelectric detector.   4

                                       Total   100

INDEX

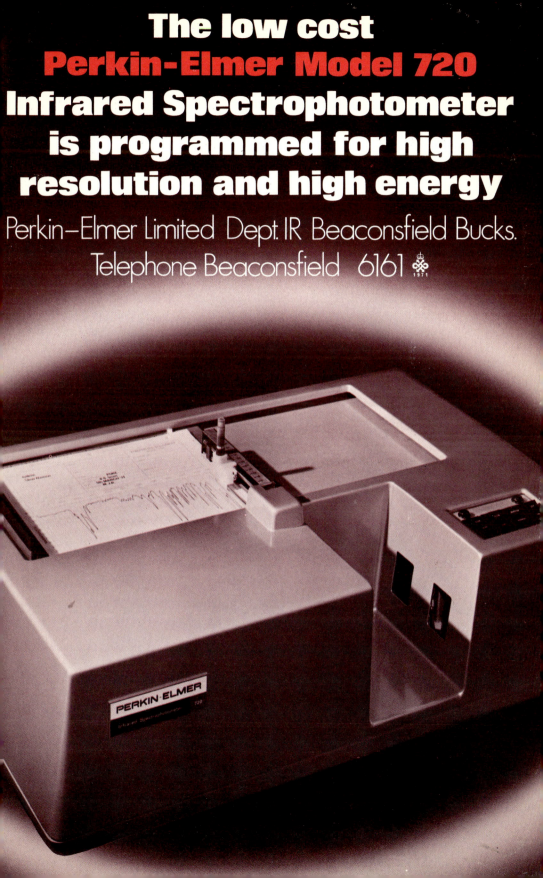

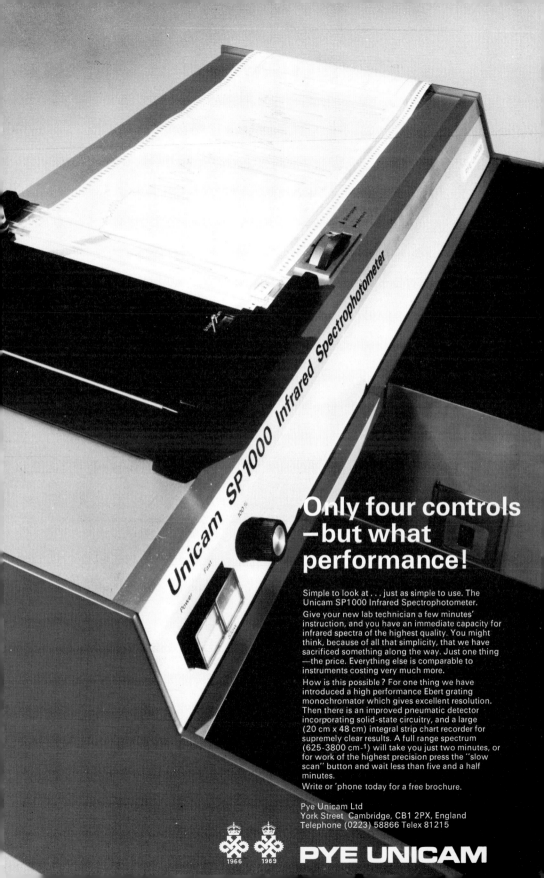

Only four controls –but what performance!

Simple to look at . . . just as simple to use. The Unicam SP1000 Infrared Spectrophotometer.

Give your new lab technician a few minutes' instruction, and you have an immediate capacity for infrared spectra of the highest quality. You might think, because of all that simplicity, that we have sacrificed something along the way. Just one thing —the price. Everything else is comparable to instruments costing very much more.

How is this possible? For one thing we have introduced a high performance Ebert grating monochromator which gives excellent resolution. Then there is an improved pneumatic detector incorporating solid-state circuitry, and a large (20 cm x 48 cm) integral strip chart recorder for supremely clear results. A full range spectrum ($625-3800$ cm^{-1}) will take you just two minutes, or for work of the highest precision press the "slow scan" button and wait less than five and a half minutes.
Write or 'phone today for a free brochure.

Pye Unicam Ltd
York Street Cambridge, CB1 2PX, England
Telephone (0223) 58866 Telex 81215

1966 1969

PYE UNICAM

Further reading in from the Heyden

Educational

BASIC INFRARED SPECTROSCOPY (2ND EDITION) 1972
J. H. van der Maas. 5 chapters. 108 pp. £1.50; $ 4.00; DM 14.00.
The author has drawn upon his vast lecturing experience in this field, to provide an authoritative guide for beginners. The theory of essential concepts and equations is covered, together with information on instrumentation, sampling, structural isomerism, quantitative analysis, etc. "It will serve as a good introduction to infrared spectroscopy" from a review in Laboratory Practice. This book is an ideal companion to the Programmed Introduction.

Theory and Practice

LABORATORY METHODS IN IR SPECTROSCOPY (2ND ED.) 1972
Edited by R. G. J. Miller and B. C. Stace.   22 chapters. £4.95; $ 13.50; DM 47.00.
The first edition, published in 1965, rapidly established itself as a standard work in the field. Chemistry & Industry, reviewing the first book, said: "...... this handbook is to be welcomed ...... a copy should be available in every infrared laboratory." See the whole page ad. for fuller details of this new vastly enlarged, up-dated edition.

HEYDEN & SON LTD